ELEMENTS IN FISH ENDOCRINOLOGY

ELEMENTS IN FISH ENDOCRINOLOGY

Dipti Nagar

Oxford Book Company
Jaipur ▪ New Delhi

Published By

Oxford Book Company

Regd. Off. "Bony Residency", Gate No. 2
Opp. Tilak Public School, Vishweswariya Nagar,
Gopalpura Road, Jaipur - 302018, India
Ph: 0141 - 2761280, Telefax: 2761381
email - oxfordbookjpr@gmail.com
website - oxfordbookcompany.com

Delhi Office: 102, 1st Floor, Satyam House,
4327/3, Ansari Road, Darya Ganj,
New Delhi - 110002, India
Ph: 011-45652440

Distributed By
KEYA INTERNATIONAL PRESS LLP
E-mail: keyainternationalpressllp@gmail.com

Published in India, 2025

ISBN : 9789391096397

Typeset by Shivangi Computers, Jaipur

Printed and bound in India by Thomson Press Ltd.,
New Delhi

Contents

Preface

This is most relevant for studies that use preserved specimens to examine aspects of the mating system or sexual development. Most often these studies rely on the status of one character, the gonad, to determine sexual phenotype or reproductive status. However, the allocation of gonads is not always a good predictor of reproductive behaviour or success. For example, the temperate marine goby Lythrypnus dalli was initially considered a simultaneous hermaphrodite based on gonadal histology. However, behavioural observations indicate that these gobies are sequential hermaphrodites that exhibit only male or female behaviour at any one time, regardless of the presence of male and female gonads. It has thus become clear that to identify alternative sexual states, we need to examine a group or suite of characters that, together, give us an unambiguous determination of sex and mode of reproduction. These characteristics include body size and coloration, gonad anatomy and physiology, genitalia (usually the genital papilla in fish), sex-specific glands (e.g., accessory gonadal structures or testicular glands), signalling structures (e.g., soundproducing mechanisms), behaviour, brain chemistry, and endocrinology. Sex- and morph-specific vocal behaviours are paralleled by a divergence in neurobiological traits ranging from the size of the sound-producing muscles to the rhythmic firing properties of vocal neurons.

The vocal organ of midshipman fish consists of a pair of sonic muscles attached to the lateral walls of their swimbladder. These muscles are the functional analog of the laryngeal and syringeal muscles of terrestrial vertebrates and apparently share a common embryonic origin from occipital head somites. A sixfold greater vocal muscle– body weight ratio in type I males, compared to type II males and females, is paralleled by a fourfold greater muscle-fiber number and fivefold greater muscle-fiber diameter. Dimorphisms in muscle fibers extend to the subcellular level, including the width of Z-lines, which represent the points of overlap of the thin actin filaments of myofibrils; the branching of the sarcoplasmic reticulum, which is the principal site for calcium exchange; the density of mitochondria; and the concentrations of metabolic enzymes.

All these traits are greater in magnitude in type I males and probably represent structural and biochemical adaptations related to the metabolic demands of humming. Each sonic muscle is innervated by a single nerve formed by branches of ventral occipital nerve roots carrying motor axons originating from paired midline vocal-sonic motor nuclei extending from the caudal hindbrain into the rostral spinal cord. Each vocal motor nucleus innervates the ipsilateral sonic muscle. Vocal motoneurons have a round to ovoid-shaped cell body and dendrites that branch throughout both sonic motor nuclei. A single unbranched axon emerges from either the soma or near the base of a primary dendrite and exits the brain via one of the occipital nerves, eventually forming synapses with sonic muscle fibers. Intracellular recording and staining studies have identified vocal pacemaker neurons that are located ventrolaterally to motoneurons.

The two neuropeptides that have been most commonly studied in the context of sexual plasticity among teleost fish, including midshipman fish are gonadotropin-releasing hormone (GnRH) and arginine vasotocin (AVT). It appears, at least for mammals, that the proper development, adult location, and physiological function of GnRH preoptic cells are necessary for adult reproductive function. Disproportionate variation in any of these parameters often leads to reproductive dysfunction. In contrast to the restrictive organization of GnRH brain nuclei required for normal sexual development in mammals, variability in the adult status of the GnRH centres in the forebrain's preoptic (POA) in fish and may play a role in the production of the wide range of sexual systems.

The book will not only meet the requirement of students but will also be useful as a guideline to the teachers and researchers.

— *Dipti Nagar*

Chapter 1

The Pineal Gland of Fishes

The pineal gland is part of the endocrine system involved with rhythmic activity in e.g. fish. The main product of the pineal gland is the indole hormone melatonin, synthesised from the amino acid tryptophan. Melatonin is mainly synthesized when it is dark, as light inhibits production. For this reason melatonin is thought to be strongly involved in biological rhythms. The pineal gland is a photoneuroendocrine gland which secretes the hormone melatonin and conveys information to the brain via neural pathway. In fish the entire system (the photodetector, the circadian clock and melatonin synthesizing enzymes) is located in the pineal organ. Several experimental studies indicated that the pineal organ is able to translate environmental information (photoperiod and temperature) into rhythmic messages and the pineal hormone melatonin is the internal chemical messenger of environmental signal or *Zeitgeber* and controls a number of functions, especially reproduction, in vertebrates.

The pineal gland is considered to be part of the system regulating biological rhythmicity, mainly due to its main secretory product, the indole hormone melatonin. In birds and mammals melatonin is strongly involved in the synchronization of diurnal and annual rhythms.

The pineal organ consists of the pineal gland and the parapineal organ. The parapineal organ is present in embryologic stages of development but is lacking in adult fish. Whereas the role of melatonin in fish is less clear. Three types of cells are considered the main content of the pineal gland, i.e. pinealocytes (photoreceptor cells), glial (supporting) cells and second order neurons (ganglion cells;. There are blood vessels supplying blood to all parts of the pineal gland, but they do not penetrate into the parenchyma of the gland. The pinealocytes are both photosensitive, containing photopigments, and secretory, producing chemical substances. Pinealocytes have been shown to undergo morphological changes in response to changes in photoperiod. The photoreceptors of the pinealocytes have similarities with the cone photoreceptors of the retina. They differ, however, as the time it takes for a resting membrane potential (the dark potential) in the individual pineal photoreceptors to reach peakpotential (hyperpolarizrtion) and the recovery time (the return to the resting potential after a stimulus) is much longer in the pineal gland than in the retina). This supports the theory that the pineal gland records gradual light intensity changes rather than the rapid changes that the retina can perceive. Further, the pineal pigments absorb light at longer wavelengths than the retina. Photoreceptors of lower vertebrates presumably release neurotransmitters constantly during darkness, and the size and numbers of synaptic ribbons (part of the photoreceptor axon terminal) change with time of day and ambient light.

Melatonin (N-acetyl-5-methoxytryptamine) is an indole hormone produced from the amino acid tryptophan. The biosynthesis of melatonin starts with conversion of tryptophan into 5-hydroxytryptophan by the enzyme tryptophan hydroxylase (TPOH). Hydroxytryptophan is then decarboxylated by the aromatic amino acid decarboxylase to serotonin. Arylalkylamine N-acetyltransferase (AANAT) converts serotonin into Nacetylserotonin and hydroxyindole-0-methyltransferase (HIOMT) methylates Nacetylserotonin to melatonin.

The pineal gland is the main organ for melatonin production in fish, but there is melatonin production in the retina of the eye as well. In mammals and birds melatonin is also produced in the gastrointestinal tract, and under specific nutrition-dependent circumstances this production may contribute to the levels of circulating melatonin, albeit not in a rhythmical fashion. No consistent information has been found for melatonin production outside the retina and pineal gland in fish but pinealectomized trout have shown increased mid-scotophase plasma melatonin level found that pinealectomy in goldfish did not completely abolish plasma melatonin levels. The production from the retina is considered to be mainly for local (paracrine) use and is not in phase with the plasma melatonin level.

In fish the melatonin diffuses into the blood stream directly after the synthesis. Melatonin is believed to be involved in behavioural, physiological and biochemical rhythmic activity. An indication of this is the diurnal variations in indole compounds (serotonin, 5-hydroxyindolacetic acid, 5-hydroxytryptophol and melatonin) in the pineal and that the melatonin levels are consistently higher during night time. Light inhibits the production of melatonin, while darkness removes this inhibition. The limiting factor is AANAT, which show cyclic activity with higher activity in darkness (in pike, Exos lucius, Abundance of AANAT messenger RNA transcripts varies in a manner parallel to the enzyme activity. Both the duration of the elevated nighttime level and the amplitude of plasma melatonin rhythm change in a fashion consistent with the seasonal change in photoperiod providing calendar information to the animal. It has been shown that Arctic charr maintains diel and seasonal melatonin rhythm perfectly reflecting the daylength even in lakes covered by ice and snow.

A very high night-time plasma melatonin level was recorded in these fish in September, when the lake temperature was high (~10°C), whereas in June when there is constant light at high latitudes the 24 hour plasma melatonin levels in Arctic charr were constantly under the detection limit of the assay. Diel fluctuations of pineal melatonin production have also been described in species closely related to the Arctic charr such as rainbow trout (Oncorhynchus mykiss; and Atlantic salmon as well as in many other fish. Portar, with his associates in 1995 and 1996 reported that melatonin levels decrease in pinealectomized fish.

It was demonstrated that the pineal gland was the only organ in fish, responsible for the presence of melatonin in the blood and the level oscillated regularly over 24 hours showing low values during day and high over night. A pineal control of gonadal maturation has been shown in fishes either by pinealectomy. The above authors have reported the relation between pineal gland and melatonin with reproduction.

The pineal gland controls reproduction through secretion of its hormone melatonin through pineal-hypophysispituitary-gonadal axis. Administration of melatonin inhibits the stimulatory effect of a long photoperiod and high temperature on the ovary in early preparatory phase. Treatment with melatonin during preparatory phase resulted in decreased ovarian weight and arrested ovarian recrudescence. When the goldfishes were pinealectomized in spring and exposed to long photoperiod conditions, the ovaries regressed and plasma gonadotropin levels were significantly depressed compared to sham operated animals. Sham operated goldfish exposed to short photoperiod conditions in spring had regressing ovaries whereas pinealectomized animals under this regime either spawned or had ovaries in the late vitellogenic phase. Several authors reported that gonadal function (wt. and histology) are inhibited with the administration of melatonin. It appears that pineal gland and melatonin are having inhibitory effect to thyroid hormone in fishes during gonadal development and maturation which is specially required for the sex steroidogenesis in the process of reproduction.

The pineal gland regulates carbohydrate metabolism by altering insulin responsiveness in the animal like goldfish. Pinealectomy in this species causes a decrease in liver glycogen stores and disappearance in plasma glucose. These effects occur independently of photoperiod acclimation and are seasonal in nature. The hormone from this gland was observed having a h5^ oglycemic effect in the above species. Several authors have reported that this gland and its hormone have a role to play in the colour change mechanism in fishes.

Melatonin is important in controlling the reproductive seasonality by stimulating the final stages of sexual maturation and by synchronizing the oocytematurity with optimal timing of spawning. Melatonin has also been found to affect estradiol levels in mature carp females and to indirectly influence the GtH II secretion *via* hypothalamic stimulatory (GnRH) centres.

Endocrine System - Endocrine Glands and Hormones

Endocrine System

The endocrine system consists of a number of glands located throughout the body that produce and secrete sub-stances called hormones into the blood. A hormone is a chemical that influences or controls the activity of a specific tissue or organ. Like the nervous system, the endocrine system coordinates the functioning of body parts. The nervous system is fast acting and utilises nerve impulses that travel rapidly along nerve fibres. The endocrine system is slower acting because it takes time to deliver hormones to target organs, but the action of hormones is longer lasting.

Hormone Action

Hormones can be divided into two categories: (1) peptides (used here to include amino acids, polypeptides, and pro-tein hormones) and (2) steroids (complex rings of carbon and hydrogen atoms). Peptide hormones act as messengers that stimulate cells by binding to specific receptors in the plasma membrane. The hormone receptor complex activates an enzyme that produces cyclic AMP (cAMP), a compound derived from ATP. Cyclic AMP, serving as a so-called second messenger, then activates other enzymes that carry out various cellular activities. Other peptide hormones utilise calcium (Ca2+) as a second messenger. In contrast to a two-messenger system, steroid hormones (which are lipid-soluble) pass through the plasma membrane and bind to receptors present in the cytoplasm. This hormone receptor complex moves into the nucleus, where the complex activates certain genes involved in protein synthesis.

Endocrine Glands

All hormones are carried throughout the body by the blood, but each one affects only a specific body part or parts, appropriately termed the target organ(s). Endocrine glands are ductless glands that secrete hormones directly into the blood. Exocrine glands, on the other hand, secrete their products into ducts that transport them to their destination. For example, the salivary glands secrete saliva into ducts that enter the mouth.

Hormones affect metabolism, appearance, and behaviour. They help maintain homeostasis by keeping the fluid, electrolyte, and glucose balance of the blood within nor-mal limits. Hormone secretion by endocrine glands is of-ten regulated by negative feedback.

Negative Feedback

In a self-regulating negative feedback mechanism, an adaptive response dampens or even cancels the stimulus that brought about the response. An endocrine gland can be sensitive to either the condition it is regulating or to the blood level of the hormone it is producing. For example, when blood is concentrated, the hypothalamus produces a hormone that causes blood dilution, which then is a signal to the hypothalamus to stop producing the hormone. On the other hand, the pituitary gland produces a hormone that stimulates the thyroid gland.

When the blood level of a hormone produced by the thyroid rises, the pituitary gland no longer stimulates the thyroid gland.

Pancreatic Hormone

Pancreatic hormones (PP) are peptides synthesized in pancreatic islets of Langerhans, which acts as a regulator of pancreatic and gastrointestinal functions.

The hormone is produced as a larger propeptide, which is enzymatically cleaved to yield the mature active peptide: this is 36 amino acids in length and has an amidated C terminus. The hormone has a globular structure, residues 2-8 forming a left-handed polyproline-II-like helix, residues 9-13 a beta turn, and 14-32 an alpha-helix, held close to the first helix by hydrophobic interactions. Unlike glucagon, another peptide hormone, the structure of pancreatic peptide is preserved in aqueous solution. Both N and C termini are required for activity: receptor binding and activation functions may reside in the N and C termini respectively.

Subfamilies

- Pancreatic hormone subtype *IPR015480*
- Peptide YY *IPR015481*

Peptide Transport and Animal Growth: The Fish Paradigm

The modern mode of thinking that digested protein is transported from the intestinal lumen into enterocytes and then into the circulation both as amino acids (AAs) and di/tripeptides dates back to the mid-1970s. Today, we know that di/tripeptides are taken up by a single transporter known as PEPtide Transporter 1 (PEPT1). However, some questions remain unanswered. For instance, although PEPT1 knock-out models have been generated, it is not clear whether the intestinal peptide transporter plays a role in the up- or downregulation of animal growth. Indeed, by deleting the intestinal peptide transporter *pept1* (previously known as *pep-2*) body size is reduced in an invertebrate animal model, such as the nematode *Caenorhabditis elegans*, whereas *PepT1* null mice grow to normal size, body weight and organ weight. In recent years, interest in piscine PEPT1 has grown rapidly. This is owing to the recurring observation that fish can efficiently use dietary di/tripeptides to sustain development, growth and metabolism.

Major, Specific Aspects of Fish Growth

In most fish, growth is indeterminate and the majority of absorbed nutrients are invested in accreting muscle tissue. Muscle growth in juvenile fish increases both the size (hypertrophia) and number of muscle fibres (hyperplasia). By contrast, the increase in the number of fibres in mammals and birds is arrested shortly after embryonic development has been completed, and further muscle growth is mainly caused by hypertrophy of existing fibres. Skeletal muscle forms 40–60% of the body mass in most adult fish. In muscle, all of the synthesized protein accumulates as growth, unlike in other tissues, such as liver or gills, in which high fractional rates of protein synthesis are matched by similarly high rates of protein degradation, with the effect of an inherently low rate of growth efficiency. As a result, the contribution of these tissues to overall growth of the fish is minute. Since muscle may account for more than half of the fish's body mass, and muscle is characterised by low fractional rates of protein synthesis, yet the highest growth rate efficiency of any tissue, it is not surprising that muscle protein synthesis *per se* is an excellent overall measure of fish growth.

The natural life cycle of many fish species includes a fasting period as a consequence of temporal and spatial food availability in the aquatic environment. Fish are well adapted to survive for long periods without food, and metabolic depression seems to be an important strategy in response to periods of food scarcity . When food levels are restored, growth can increase over and above normal rates in these fish. This phenomenon is known as compensatory growth. The fasting/refeeding paradigm has been widely used to investigate the molecular mechanisms underlying the transition from a catabolic to an anabolic state in fish. For instance, in juvenile rainbow trout (*Oncorhynchus mykiss*) fasted for 10 weeks, refeeding resulted in a 15-fold increase in insulin-like growth factor I (IGF-I) mRNA 12 days following refeeding, and in a much smaller increase in the less abundant insulin-like growth factor II (IGF-II) mRNA.

Peptide Transport(Ers) and Growth: the Fish Paradigm

Recently, sea bass (*Dicentrarchus labrax*) has emerged as an excellent fish model for muscle growth studies as it exhibits extensive juvenile (post-metamorphosis) muscle hyperplasia, which contributes to its large adult size. We have demonstrated in this species that nutritional status influences the expression of several genes that encode for physiologically relevant proteins which participate in both an endocrine and autocrine/paracrine fashion in promoting compensatory growth induced by refeeding. In our experiments, sea bass nutritional status influenced the expression of key growth-related genes, such as myostatin in muscle (electronic supplementary material), IGF-I and IGF-II in liver and muscle (electronic supplementary material), and these regulatory responses correlated highly with changes in fish body weight and condition factor (a morphological indicator of body shape). At the gastrointestinal level, the stomach responded to fasting by upregulating ghrelin (electronic supplementary material) and downregulating gastricsin (electronic supplementary material) expression as part of an integrated response to food availability. Under the same experimental conditions, PEPT1 expression in the proximal intestine decreased during fasting and increased during refeeding, which is the opposite of what is observed in mammals and birds, suggesting that PEPT1 operates in direct support of fish (muscle) growth.

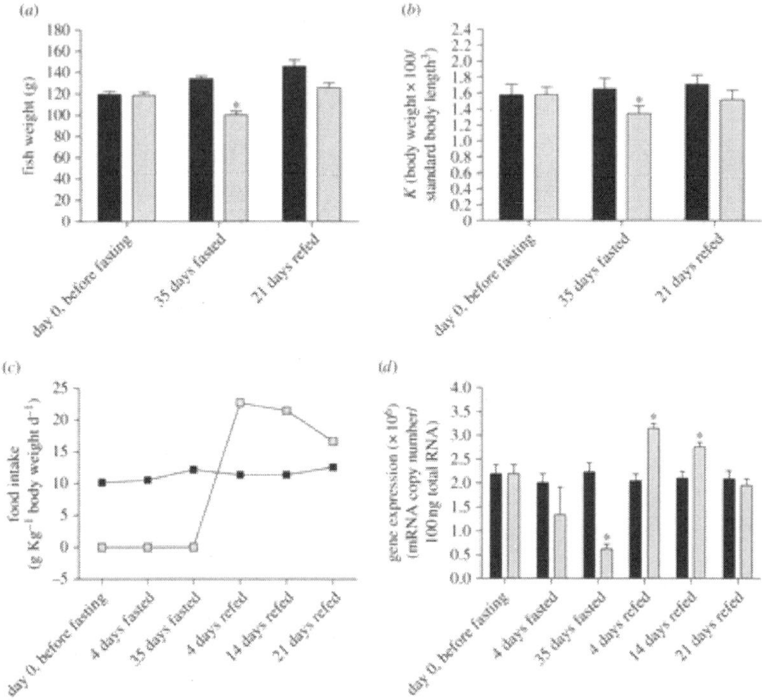

Figure: (*a*) Mean body weight and (*b*) condition factor (K = body weight × 100/ standard body length) of sea bass (*Dicentrarchus labrax*) control (black bars) and fasted and refed group (grey bars). Fish of the control (fed ad libitum) and treated (fasted and refed) group were weighted before fasting (day 0, before fasting), at the end of fasting (35 days fasted) and then at 21 days following refeeding (21 days refed). (*c*) Weekly changes in mean daily food intake in the group of sea bass subjected to fasting and refeeding (grey squares) and in that fed ad libitum (black squares). (*d*) Expression levels of PEPT1 in proximal intestine, measured by real-time PCR. PEPT1 mRNA copy number was normalized as a ratio to 100 ng total RNA. Fish were sampled before fasting (day 0, before fasting), 4 days after fasting (4 days fasted), at the end of fasting (35 days fasted), and then sequentially at 4, 14 and 21 days following refeeding (4 days refed, 14 days refed and 21 days refed). Black bars represent fed ad libitum and grey bars represent fasted and refed. Means ± s.e.m. of five animals in each group are shown. Asterisks (*) indicate significant differences between treated and control groups for each time point tested.

These data document a correlation between the expression of an intestinal transporter, such as PEPT1, which is primarily involved in the uptake of dietary protein degradation products, and animal growth for the first time, to our knowledge, in a vertebrate model. In fish, PEPT1 provides the necessary amount of AA to sustain net muscle protein synthesis, and thus body growth.

Further evidence indicating that (synthetic) dietary dipeptide containing indispensable AA is responsible for enhanced weight gain of fish is supported by the significantly increased PEPT1 expression in the enterocytes of salmonid fish.

Perspectives

In our opinion, fish represent an excellent model for studying the effects of protein administration on animal growth because of the direct correlation that exists between dietary protein availability and accretion of body mass.

In this respect, PEPT1 function is perfectly integrated into the physiological response scheme, representing a route for absorbing the bulk of protein that is needed to sustain body growth. Of course, this is a simplified model that cannot be applied to higher vertebrates (humans included), which are determinate growers.

In fact, PEPT1 expression increases during fasting and decreases during refeeding in mammals and birds, which suggests a different, and most probably more sophisticated and complex control of the transporter activity. However, this does not exclude the possibility that PEPT1 may play a significant role in developmental stages in which linear accretion prevails even in higher vertebrates.

Finally, an increasing number of researchers in the fish biological sciences are currently focusing on and discussing the potential role of peptide transport for the uptake of dietary AA in fish.

This attention mainly derives from the observation that teleosts can efficiently use dietary di/tripeptides for development, growth, and metabolism and, consequently, that balanced peptide-based diets or peptide rather than AA supplementation would be highly beneficial in solving the nutritional inadequacy problems of formulated feeds for cultivated fish.

Materials and Methods

Ethical Approval

This study was conducted on Atlantic salmon (Salmo salar L.) of aquaculture origin. Fish were humanely killed by a blow to the head following Schedule 1 of the Animals (Scientific Procedures) Act 1986 (Home Office Code of Practice. HMSO: London January 1997), and all husbandry procedures and experimental protocols were approved by University of St Andrews Animal Ethics and Welfare Committee.

Fasting-Feeding Experiments

Two independent experiments were performed to manipulate growth.

The first experiment was used to obtain samples for preparation of the subtractive libraries and has been described previously (13). A strain of Norwegian farmed salmon (Salmo Breed) was used comprising 110 fish, 840 ± 152 g, mean ± SD. Fish were reared in replicate tanks under continuous light at an average temperature of 7.8°C, 13.96 mg/l oxygen, and average salinity 28.9 ppm at Ewos Innovation (Dirdal, Norway).

The salmon were individually passive integrated transponder tagged (Fish Eagle, Lechlade, Gloucestershire, UK) to monitor growth and fed a maintenance diet (25% normal ration) to achieve a zero or slightly negative growth rate over a period of 22 days. Growth was measured as the thermal growth coefficient (TGC) = $[(M_2^{0.333}$ “ $M^{0.333})(\text{degree days})^{"1}$ *1,000], where M_1 and M_2 were start and final body weights, respectively. Degree days values are the sum of the °C values for each day of the experiment. At the start of the experiment [0 day (d)], fish were fed to satiation with a commercial feed (EWOS Innovation) to stimulate fast growth. Three fish were sampled at 0 d (fasted, 897 ± 72.8 g, $n = 3$), and at 3 d (855 ± 99 g, $n = 3$) and 14 d (842 ± 56 g, $n = 3$) following satiation feeding. Fast muscle was dissected from the dorsal epaxial myotome between 0.6 and 0.7 standard length (tip of snout to last vertebrae). Tissues were snap-frozen in liquid nitrogen and stored at “80°C until analyzed.

A second independent experiment involving a change in feeding/growth status was carried out to validate the discovery of muscle growth-related genes from the first experiment with high stringency. The expression of candidate growth related genes was investigated in the context of 26 other myogenic genes with well established roles in growth regulation. By using a different population, size class, and rearing conditions of fish for validation of expression patterns we hoped to identify genes that changed with feeding and growth in a variety of contexts, i.e., were not specifically regulated at one particular life stage or set of environmental conditions. For this purpose we used a population of farmed salmon parr sourced from Scotland reared in duplicate tanks at the University of St Andrews comprising 150 fish, 59.8 ± 7.9 g, mean ± SD.

The average temperature was 10.6°C, and fish were reared in aerated freshwater under a photoperiodic regime of 12 h light:12 h dark. Fish were fed a maintenance diet (25% normal ration) for 21 days, fasted for 7 d, and fed to satiation (starting at 0 d) with a commercial feed (EWOS Innovation). Sampling of fish occurred at 0, 1, 3, 5, 8, 15, and 21 d with three fish sampled from both replicate tanks at each time point. Individual mass and fork lengths were measured. Dissection and preparation of fast skeletal muscle samples were exactly as described for the first experiment. In addition, slow muscle, heart, liver, brain, kidney, gill, skin, gut, and eye tissues were dissected from fasted fish at 0 d and from fish fed to satiation for 8 and 21 d. RNA was extracted from these tissues from three fish at each time point, pooled and cDNA synthesized as described below.

Myogenic Cell Culture

Primary cell cultures from fast skeletal muscle were performed as previously described (10) using the same population of fish and rearing conditions as described for the second growth manipulation experiment above.

RNA Extraction

Total RNA was extracted by addition of 100 mg of muscle to Lysing matrix D (Qbiogene, Irvine, CA) with 1 ml Tri Reagent (Sigma, Gillingham, Dorset, UK) as per manufacturer's recommendations and homogenized using a Fast Prep instrument (Qbiogene). Total RNA was quantified based on absorbance at 260 nm. Genomic DNA contamination was removed by treatment with Turbo DNA-free (Ambion, Austin, TX), and the integrity of purified RNA confirmed by agarose gel electrophoresis. For myogenic cell culture, RNA was immediately extracted from duplicate wells of three separate cell cultures. RNA extraction and genomic DNA removal was performed using a RNeasy plus kit (Qiagen, Chatsworth, CA) as per manufacturer's recommendations.

Subtractive Library Production

RNA was extracted and pooled from the fast skeletal muscle of three fish from each of the 0 (fasted), 3, or 14 d (satiation feeding) samples from the first growth manipulation experiment. The SMART cDNA synthesis kit (Clontech Laboratories, Mountain View, CA) was used to synthesize cDNA. SMART cDNA was amplified for 17 cycles and digested with *Rsa*I (Clontech Laboratories), and used for SSH using the PCR-select cDNA subtraction kit (Clontech Laboratories). Four subtractions were performed in total, two forward (where cDNA from *day 3* or *day 14* was used as tester and cDNA from *day 0* was used as driver), and two reverse (cDNA from *day 0* was used as tester and cDNA from *days 3* or *14* used as driver).

The subtracted PCR products were ligated into pCR4-TOPO vector (Invitrogen, Carlsbad, CA) and transformed into TOP10 chemically competent Escherichia coli cells (Invitrogen). Individual colonies were picked and grown overnight in 200 ìl of Luria broth containing 100 ìg/ml ampicillin. Four subtractive libraries were produced consisting of 3,072 clones in the *day 3* forward (3F); 3,264 clones in the *day 14* forward (14F); 1,536 clones in the *day 3* reverse (3R); and 1,920 clones in the *day 14* reverse (14R).

EST Sequencing

Inserts were amplified by transferring 1 ìl of culture to a PCR reaction. PCR was performed in a 20 ìl reaction volume using BioTaq PCR kit (Bioline, London, UK) under limiting conditions of primers and dNTPs, using short T3 (52 -ATTAACCCTCACTAAAG-32) and short T7 (52 -AATACGACTCACTATAG-32) primers. PCR was performed with an initial denaturation at 96°C for 2 min, followed by 35 cycles of amplification (96°C for 20 s, 49°C for 20 s, and 72°C for 45 s), and final extension at 72°C for 5 min.

PCR products were diluted two fold and directly sequenced using a T3 primer (52 -AATTAACCCTCACTAAAGGG-32) and Big Dye terminator sequencing mix (Applied Biosystems). The sequencing reaction cycle (95°C for 20 s and 60°C for 140 s) was repeated 25 times. DNA sequences were read using an Applied Biosystems 3700 capillary sequencer (Geneservice, Oxford, UK).

Bioinformatic Analysis

Sequence trace files were first analyzed by trace2dbEST (58) with removal of low-quality sequence, vector sequence, and contaminating E. coli sequences and polyA tails. The processed sequences were subjected to BLAST similarity searches using the BLASTX algorithm against the National Centre for Biotechnology Information (NCBI) nonredundant protein sequence database. Good-quality sequences and their preliminary annotation were submitted to dbEST (9). Clustering of sequences was performed using the Phrap algorithm within the Codon Code program (Codon Code, Dedham, MA).

Annotation of ESTs was performed with Blast2GO (23) using the default parameter settings. Annotations were assigned to three ontologies defined by the Gene Ontology (GO) consortium and were further summarised by assigning GO slim terms.

Quantitative Real-time PCR

The following procedures were performed to comply with the Minimum Information for Publication of Quantitative Real-Time PCR experiments MIQE guidelines (18). RNA was concentrated by ethanol precipitation and quantified using a NanoDrop 1000 spectrophotometre (Thermo Fisher Scientific, Waltham, MA). Only RNA with an A260/280 ratio between 1.8 and 2.1 and an A260/230 >1.9 was used for cDNA synthesis. Residual genomic DNA was removed using the genomic DNA wipeout buffer included in the Quantitect reverse transcription kit (Qiagen). We reverse transcribed 1 ìg and 800 ng of RNA from the second growth manipulation experiment and myogenic culture, respectively, into cDNA for 30 min at 42°C using a Quantitect reverse transcription kit (Qiagen) as per manufacturer's recommendations.

qPCR was performed using a Stratagene MX3005P QPCR system (Stratagene, La Jolla, CA) with Brilliant II SYBR (Stratagene). cDNA used in qPCR was first diluted 80-fold with nuclease-free H_20. Each qPCR reaction mixture contained 7.5 ìl 2× Brilliant II SYBR green master mix (Surestart Taq DNA polymerase, 2.5 mM $MgCl_2$), 6 ìl cDNA (80-fold dilution), 500 nM each primer and RNase-free water to a final volume of 15 ìl. Amplification was performed in duplicate in 96 well plates (Stratagene) with the following thermal cycling conditions: initial activation 95°C for 10 min, followed by 40 cycles of 15 s at 95°C, 30 s at 60°C, and 30 s at 72°C. Control reactions included a no template control and no reverse transcription control. Dissociation analysis of the PCR products was performed by running a gradient from 60 to 95°C to confirm the presence of a single PCR product. Products were also sequenced to confirm identity. The PCR amplification efficiency of each primer pair was calculated with LingregPCR 2009 using the average efficiency of 92 PCR reactions for each gene (67).

Primer Design

Primers for reference genes (10) and for 25 other genes known to be involved in myogenesis and growth (11–14) have been previously published as have primers for *CTSL1* (88) and *FXYD8* (82). Primers were designed using Primer 3 (65) to have Tm of 60°C, and where possible, were designed to cross an exon-exon junction to avoid amplification of contaminating genomic DNA. To determine exon-intron junction sites, genomic sequences for orthologous genes from Danio rario, Gasterosteus aculeatus, Oryzias latipes, Takifugu rubripes, and Tetraodon nigroviridis were retrieved from Ensembl and compared with the *Salmo salar* cDNA sequences using the Spidey software tool. Primers were designed across conserved exon-intron junctions and used at a final concentration of 500 nM.

QPCR Data Analysis

GeNorm (84) was used to analyze the stability of transcription of several reference genes including hypoxanthine phosphoribosyltransferase 1 (*HPRT1*), â-actin, RNA polymerase II, elongation factor 1 alpha (*ef1*á), ribosomal protein s29 (*rps29*), and ribosomal protein L13 (*rpl13*). GeNorm Analysis revealed *rps29* and *rpl13* to be the most stable genes in the refeeding experiment (geNorm stability value M = 0.19) and also across the different tissues. Normalization of gene expression was performed using the geometric average of *rpl13* and *rps29*, and values are shown as arbitrary units. Statistical analysis was performed using minitab (Minitab). Normalization of gene expression in myogenic cell culture was performed using the geometric average of *ef1*á, *PPIA*, and *HPRT1* as described by Bower et al.. Significant differences in expression between time points were calculated by ANOVA using Fisher's individual error rate post hoc tests. Correlations in gene expression were calculated using Spearman rank correlation. Hierarchical clustering analysis of z-score normalized gene expression data was performed with Permutmatrix.

First Growth Manipulation Experiment

CDNA Library Characterisation

To identify genes that were upregulated in fish that showed no net growth or slightly negative growth and that responded early and late following satiation feeding to stimulate fast growth, a total of four subtractive libraries were produced, with forward and reverse subtractions performed to compare 0 d to 3 d, and 0 d to 14 d. Fish at 0 d had a zero or slightly negative growth rate (TGC = "0.26 ± 0.4, mean ± SD, n = 10), while after 14 d satiation feeding growth rate was positive (TGC = 1.4 ± 0.3, mean ± SD, n = 10). Forward and reverse subtractions were performed to compare 0 d to 3 d, and 0 d to 14 d groups.

Forward subtracted libraries (3F and 14F) were enriched for transcripts that increased in response to feeding, while the reverse libraries (3R and 14R) were enriched for transcripts that decreased in response to feeding. A total of 10,000 clones were single-pass sequenced giving a total of 7,320 good quality ESTs (2,330 from 3F, 2,421 from 14F, 1,173 from 3R, and 1,396 from 14R). These ESTs, with an average read length of 567 bases, were submitted to dbEST under the accession numbers GE617488–GE624934.

Sequence assembly using Phrap identified 222 contigs and 490 singletons in the 3F library, 242 contigs and 530 singletons in the 14F library, 133 contigs and 200 singletons in the 3R library, and 162 contigs and 206 singletons in the 14R library. The largest clusters present in the libraries comprised muscle structural proteins such as troponin I and actin, and metabolic genes such as glycogen phosphorylase, triose phosphate isomerase and glyceraldehyde 3 phosphate dehydrogenase; however, in the 3F library, two large clusters were present comprising 99 and 100 ESTs that corresponded to *HSP90A*, and a third cluster comprising 31 ESTs corresponding to *HSPA1B*.

To identify sequences that were specific to a particular library, the nonredundant sequences (i.e., consensus sequences from contigs and singletons) from each library were aligned. Within the reverse libraries (3R + 14R), there were 82 nonredundant sequences that formed contigs and 523 singletons (data not shown). A total of 238 nonredundant sequences formed contigs from the 3F and 14F libraries, 173 formed contigs from the 3F and reverse libraries, 171 formed contigs from the 14F and reverse libraries, and 196 formed contigs from 3F, 14F and 3R + 14R libraries. The contigs mainly comprised muscle structural proteins and metabolic enzymes.

Many of the contigs from each library did not have significant hits against the NCBI nonredundant protein database (51% of 3F, 48% of 14F, 42% of 3F, 48% of 14R) and are therefore novel sequences or untranslated regions. The nonredundant collection of singletons and contigs that did have significant hits were annotated and assigned to three ontologies defined by the GO consortium and further summarised by assigning GO slim terms. There were 72, 75, 38, and 39 contigs from the 3F, 14F, 3R, and 14R libraries, respectively, that did not have an associated GO term and are therefore unclassified.

To identify processes and genes that were differentially regulated between the libraries, individual ESTs were annotated using Blast2GO (23), and the relative abundance of the GO terms in the forward and reverse libraries compared using the GOSSIP software package within Blast2GO using a false discovery rate P-value cutoff of $P < 0.01$ (7). Based on the GO-term distribution, in the 3F library, genes associated with vesicle formation, nucleotide binding, developmental processes, cytoskeleton organisation, response to stimulus, and cell differentiation were enriched compared with the 3R library.

In the 3R library, genes with GO terms associated with cell death and transferase activity were enriched compared with the 3F library. Comparing the GO term distribution for the 14F and 14R libraries revealed genes associated with ion channel activity, ion binding, vesicle formation, and transporter activity were enriched in the 14F library, while those associated with macromolecular complex and cellular process were enriched in the 14R library. The GO-term distribution suggests that there are more genes differentially expressed at *day 3* than at *day 14* of feeding. Many of the GO terms that were differentially distributed between forward and reverse libraries corresponded to heat shock proteins and in particular *HSP90A*.

Selection of Candidate Genes

Transcripts putatively identified as being nutritionally regulated and growth related are referred to as candidate genes throughout the rest of the paper. The results obtained from the interlibrary assembly of nonredundant sequences and the GO term enrichment analysis were used to make an initial selection of candidate genes based on the following criteria: *1*) Contigs that contained ESTs present in the forward library and absent from the reverse (or vice versa) and *2*) contigs significantly enriched in the forward or reverse library based on the relative abundance of GO terms. These selection criteria identified differentially expressed genes of known function as well as genes with no associated GO term and allowed for the selection of a subset of 40 genes for detailed investigation including abundant transcripts, transcription factors, and signalling molecules. We selected 30 genes putatively upregulated with feeding from the forward libraries along with 10 genes putatively downregulated with feeding from reverse libraries.

For ease of interpretation, throughout results and discussion, when homology between the candidate gene and top blastx hit were high (blastx e-value <1e-3) we will refer to the candidate genes using the gene symbol from their corresponding top blastx hit. When there is poor or no homology between candidate gene and known genes (blastx e-value >1e-3), we will refer to the candidate gene by their contig number.

Chapter 2

Gene Expression Profiling of Candidate Genes

Second Growth Manipulation Experiment

The aim of this experiment was to determine which of the selected gene candidates were nutritionally regulated and associated with a change in growth status over a broad range of physiological contexts.

To this end we used a different population of Atlantic salmon from the first experiment, investigated a juvenile freshwater rather than adult seawater stage, and maintained the fish under very different environmental conditions.

To obtain an insight into the function of the candidates, their expression following fasting and refeeding was clustered in relation to the expression of 26 other genes with well-characterised roles in myogenesis and growth regulation. These 26 genes were examined in the same cDNA, and the data processed identically to that of the candidate genes.

Expression of several other candidate genes was also modulated with feeding in the second experiment; however, the changes in expression were not statistically significant, and these were discarded from further analysis. Apart from the expression of *HSP30* (which was expected to be upregulated but was actually downregulated), the candidate genes gave the expected expression profiles as predicted by the selection criteria.

Genes Upregulated During Feeding

The upregulated genes showing the greatest fold change with feeding in the second growth manipulation experiment were *RTN1* (21-fold), *FCon14* (19-fold), *HSPA1B* (9-fold), and *HSP90A* (5-fold).

There were distinct phases of gene expression observed during the feeding response.

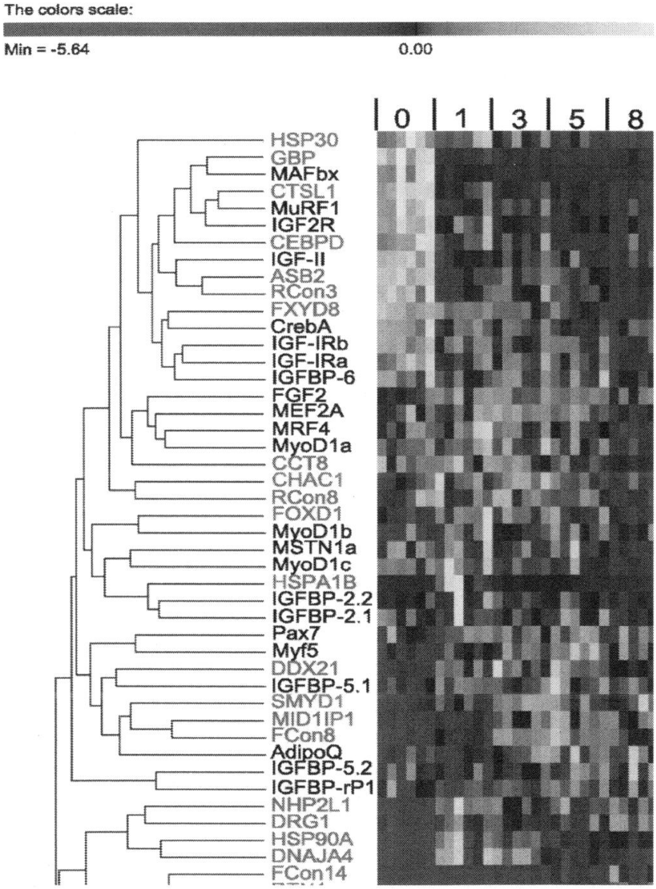

Figure: Heat map summary and hierarchical clustering for candidate genes (red) with other genes involved in muscle growth regulation during refeeding of fasted Atlantic salmon [0 days (d)] and fish fed to satiation for 1, 3, 5, 8, 15, and 21 d. Rows are standardized to have mean 0 and standard deviation (SD) 1 so that yellow indicates high and blue indicates low values.

For example, genes which responded early to refeeding included *HSPAIB* at 1 d and *CHAC1* and *FOXD1* at 1 d–5 d. *FOXD1* and *HSPA1B* expression clustered with *myoD1b*, *myoD1a*, and the *IGFBP-2* paralogs, respectively. *FCon8* and *MID1IP1* expression was correlated (R = 0.80, $P < 0.001$) and had highest levels at 3 and 5 d, which then remained elevated throughout the experiment ($P < 0.05$). This pattern was similar to that of *SMYD1* and *DDX21*, and together the expression of these genes formed a cluster that included *IGFBP-5.1*, *AdipoQ*, *Pax7*, and *Myf5*.

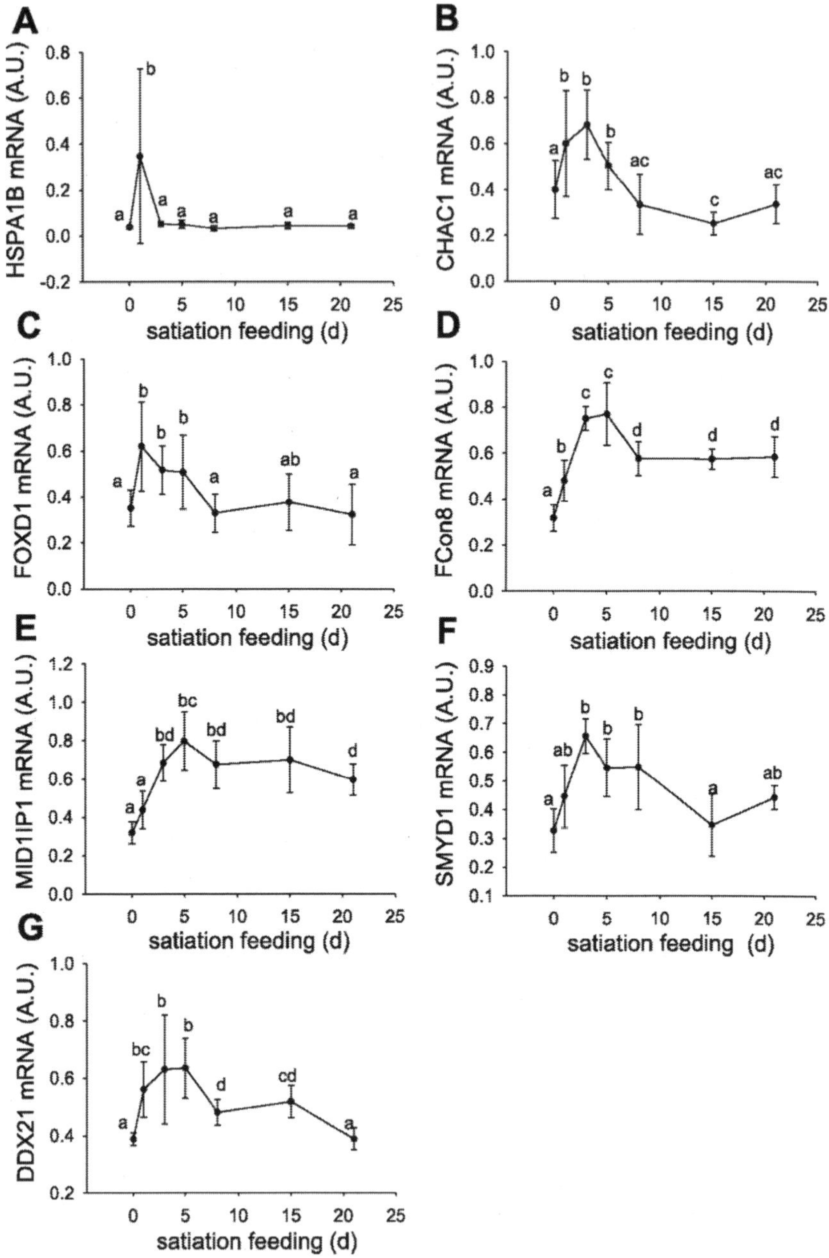

Figure: Expression profiles for *HSPA1B* (*A*), *CHAC1* (*B*), *FOXD1*(*C*), *FCon8* (*D*), *MID1IP1* (*E*), *SMYD1* (*F*), and *DDX21* (*G*) in fasted fish (**0** d) and fish fed to satiation for 1, 3, 5, 8, 15, and 21 d. Values represent means ± SD, 6 samples per point. Significant differences between means are indicated by different letters.

Another group of candidate genes displayed more sustained increase in expression after feeding. For example, *DNAJA4* , *DRG1*, *HSP90A*, and *NHP2L1* all showed peak expression within 1 d of refeeding ($P < 0.05$), which, with the exception of *DRG1* and *NHP2L1* at 21 d, remained above the levels observed in fasted fish ($P < 0.05$). *RTN1* and *FCon14* showed a monotonic increase in expression throughout refeeding, similar to that of *FCon5*, and had peak expression at 21 d ($P < 0.05$).

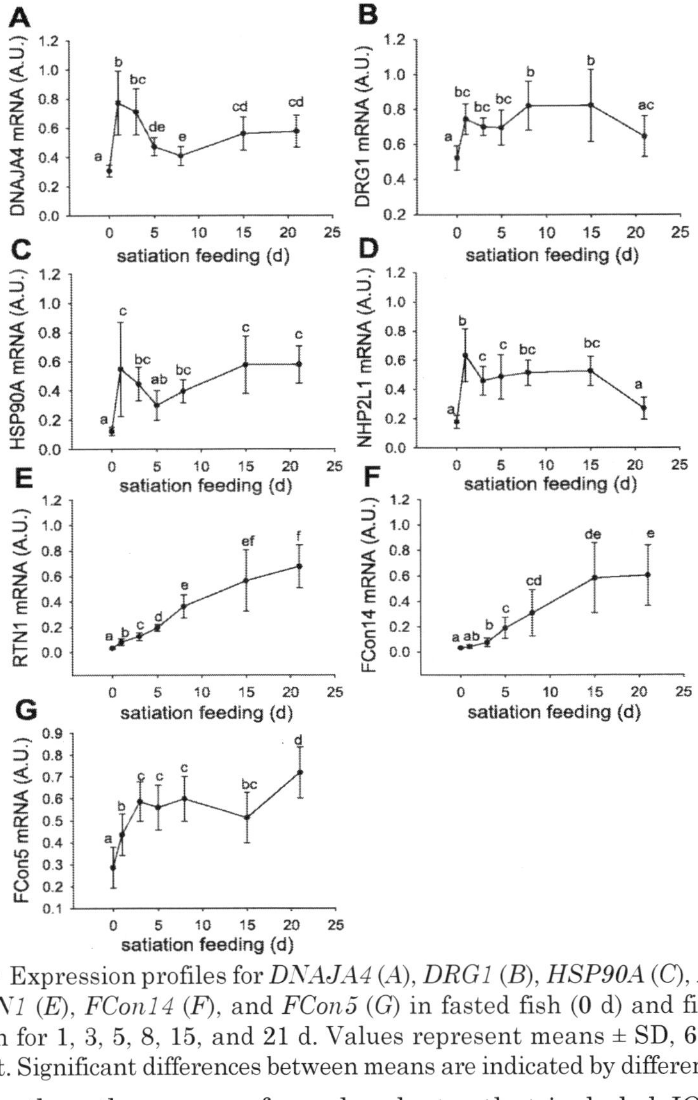

Figure: Expression profiles for *DNAJA4* (*A*), *DRG1* (*B*), *HSP90A* (*C*), *NHP2L1* (*D*), *RTN1* (*E*), *FCon14* (*F*), and *FCon5* (*G*) in fasted fish (0 d) and fish fed to satiation for 1, 3, 5, 8, 15, and 21 d. Values represent means ± SD, 6 samples per point. Significant differences between means are indicated by different letters.

Together, these genes formed a cluster that included *IGF-I* and *IGFBP-4*, with expression of several genes correlated including

DNAJA4 and *HSP90A* (R = 0.74, $P < 0.001$), *RTN1* and *FCon14* (R = 0.81, $P < 0.001$), *RTN1* and *IGF-I* (R = 0.82, $P < 0.001$) and *FCon14* and *IGF-I* (R = 0.75, $P < 0.001$).

Genes Downregulated During Feeding

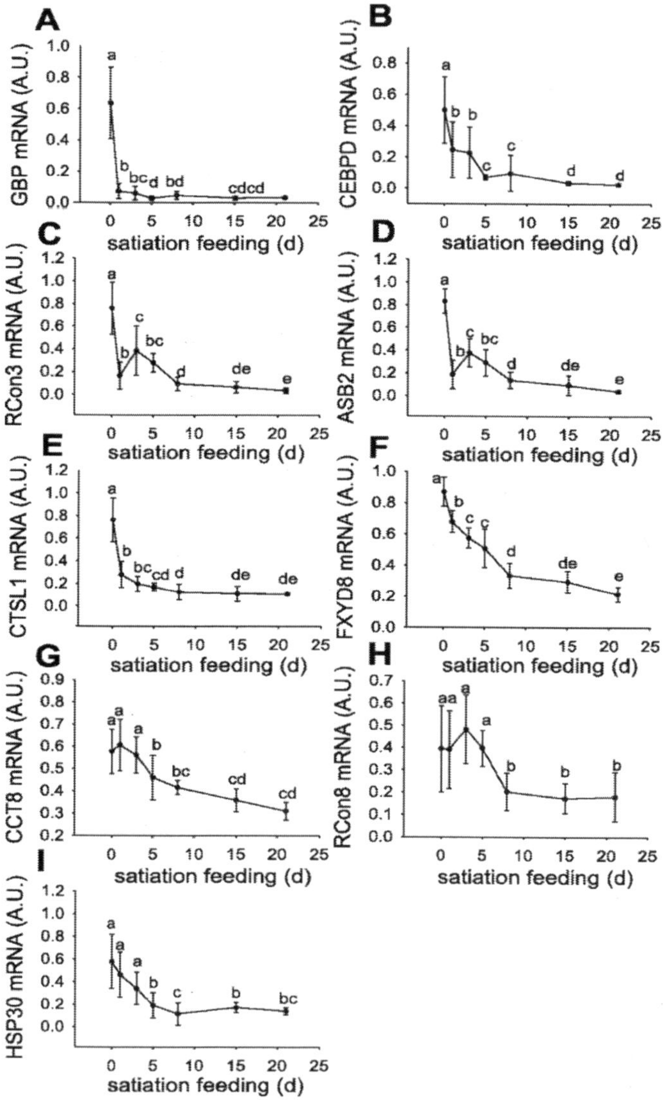

Figure: Expression profiles for *GBP* (*A*), *CEBPD* (*B*), *RCon3* (*C*), *ASB2* (*D*), *CTSL1* (*E*), *FXYD8* (*F*), *CCT8* (*G*), *RCon8* (*H*), and *HSP30* (*I*) in fasted fish (0 d) and fish fed to satiation for 1, 3, 5, 8, 15, and 21 d. Values represent means ± SD, 6 samples per point. Significant differences between means are indicated by different letters.

Of the downregulated genes, those showing the greatest fold change were *GBP*, *CEBPD*, *RCon3*, *ASB2*, *CTSL1*, and *FXYD8*. *GBP*, *CEBPD*, *RCon3*, *ASB2*, *CTSL1*, and *FXYD8* were all strongly downregulated as early as 1 d following refeeding ($P < 0.05$) and remained at this depressed level of expression throughout the experiment. The expression of these genes clustered and was highly correlated with *MAFbx* and *MuRF1* genes known to be involved in protein degradation. For example, *GBP* and *CEBPD* were highly correlated with *MAFbx* ($R = 0.93$, $P < 0.001$ and $R = 0.76$, $P < 0.001$, respectively), while *CTSL1* had high correlation with *MAFbx* ($R = 0.88$, $P < 0.001$) and *MuRF1* ($R = 0.95$, $P < 0.001$). *ASB2* and *RCon3* expression was highly correlated with each other ($R = 0.91$, $P < 0.001$) and also with that of *IGF-II* ($R = 0.81$, $P < 0.001$ and $R = 0.81$, $P < 0.001$, respectively). Expression of *CCT8*, RCon8, and *HSP30* showed a slower response, being significantly downregulated from 8 and 5 d ($P < 0.05$).

Tissue Distribution of Candidate Genes

To further characterise the candidate genes and to identify genes with high expression in muscle, we examined their expression across several tissues in fasted fish and fish fed to satiation for 8 and 21 d. The relative abundance of transcripts and their response to refeeding in various tissues were examined using QPCR. Twelve of the candidate genes had high expression in skeletal muscle compared with other tissues and included *SMYD1, RTN1, MID1IP1, HSP90A, FCon5, FOXD1, FCon8, DNAJA4, HSP30, CEBPD, ASB2, RCon3*, and *FXYD8*. Many of these genes gave the same response to refeeding in slow muscle as in fast muscle. The remaining genes were expressed at similar levels across the tissues, although *RTN1* had high expression in the brain and *RCon8* had high expression in gut.

Expression of Candidate Genes During Myogenesis in Cell Culture

To further characterise the candidate genes and investigate their possible role in myogenesis, we examined their expression profile in vitro during the maturation of a primary cell culture. Primary cell cultures derived from fast muscle of Atlantic salmon showed distinct phases of maturation. Mononuclear cells and proliferating cells predominated at 2–5 d, after which short myotubes became evident. Myotube elongation subsequently occurred resulting in continuous sheets of multinucleated myotubes at 20 d. The expression of MRFs and components of the IGF-signalling pathway and cell proliferation rates have been previously examined using the same culture used to characterise the candidates from the subtractive library.

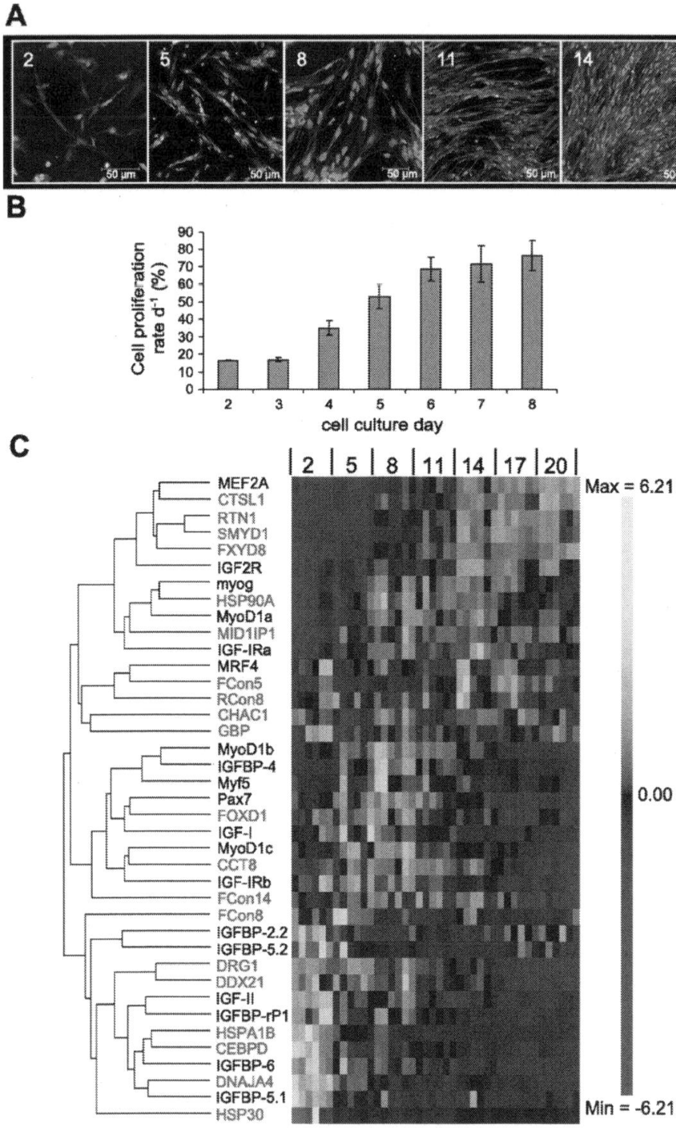

Figure: Phenotype of the cell culture at 2, 5, 8, 11, and 14 d (*A*). Cell proliferation rates of myogenic cells from 2 d to 8 d (*B*). Heat map summary and hierarchical clustering for candidate genes (red) with MRFs, *MEF2A*, *pax7*, and components of the IGF signalling pathway during Atlantic salmon myogenic cell culture at 2, 5, 8, 11, 14, 17, and 20 d (*C*). Rows are standardized to have mean 0 and SD 1 so that yellow indicates high and blue indicates low values (*C*).

These data were produced from the same cDNA samples and analyzed identically to the subtractive library candidates. To place the expression of candidate genes within the myogenic program, we combined and compared the subtractive library candidates with this data set. There was a cluster of genes that were highly expressed in *day 2* of the culture corresponding to a phase when mostly mononuclear cells were present and proliferation rates were low. These genes decreased in expression to varying extents at *day 5* and *day 8* before being downregulated during the phases of myotube elongation and differentiation. This cluster included *IGF-II*, *IGFBP-rP1*, *IGFBP-6*, *IGFBP-5.1*, as well as the candidate genes *DRG1*, *CEBPD*, *DDX21*, *HSPA1B*, *DNAJA4*, and *HSP30*.

External to, but associated with, this cluster was *FCon8*. A second major cluster of transcripts were weakly expressed at *day 2*, showed high expression at *days 5* and *8* as proliferation of myoblasts became active and short myotubes formed, but were then downregulated to low levels at later stages of differentiation when the cultures of myotubes became confluent. These transcripts were expressed at low levels at *day 2* and *day 5* of the culture but were strongly upregulated with differentiation and the formation of confluent multinucleated myotubes. Associated with the third cluster was *MRF4*, which showed a biphasic expression at early and late stages of the culture. *MRF4* formed a subcluster with the candidates *CHAC1*, *GBP*, and the novel transcripts *FCon5* and *RCon8*.

There was no significant change in expression of *FCon14* (data not shown), very low expression of *ASB2* and *RCon3* was observed in some samples at 2 d (data not shown), and *NHP2L1* expression was below the limits of detection (data not shown). Of the genes that we had previously identified as upregulated with feeding, several were also upregulated during myogenesis in vitro and included *FCon8*, *HSP90A*, *MID1IP1*, *RTN1*, *FOXD1*, *FCon5*, and *SMYD1*. These genes are therefore excellent candidates as promyogenic factors. *CEBPD* and *HSP30* were downregulated during refeeding and from 2 d in the myogenic culture and are thus excellent candidates as myogenic inhibitory genes, as their levels decreased as cells began to grow and proliferate. Several genes showed correlated expression patterns during myogenesis. Expression of *SMYD1* and *RTN1* was highly correlated ($R = 0.96$, $P < 0.001$), and also with that of *MEF2A* ($R = 0.84$, $P < 0.001$, $R = 0.84$, $P < 0.001$, respectively) and *IGF2R* ($R = 0.78$, $P < 0.001$, $R = 0.78$, $P < 0.001$, respectively). *HSP90A* and *MID1IP1* were correlated with *myoD1a* ($R = 0.84$, $P < 0.001$, and $R = 0.67$, $P < 0.001$, respectively) and *myog* (R = 0.88, $P < 0.001$ and R = 0.76, $P < 0.001$, respectively). *CTSL1* correlated with *IGF2R* (R = 0.79, $P < 0.001$) and *MEF2A* (R = 0.87, $P < 0.001$), and *FXYD8* was also correlated with *MEF2A* (R = 0.87, $P < 0.001$). *CEBPD* clustered with *IGF-II* (R = 0.75, $P < 0.001$) and *IGFBP-6* (R = 0.81, $P < 0.001$), *HSPA1B* with *IGFBP-6* (R = 0.76, $P < 0.001$), *FCon5* with *MRF4* (R = 0.68, $P < 0.001$) and *DNAJA4* with *IGFBP-5.1* (R = 0.81, $P < 0.001$).

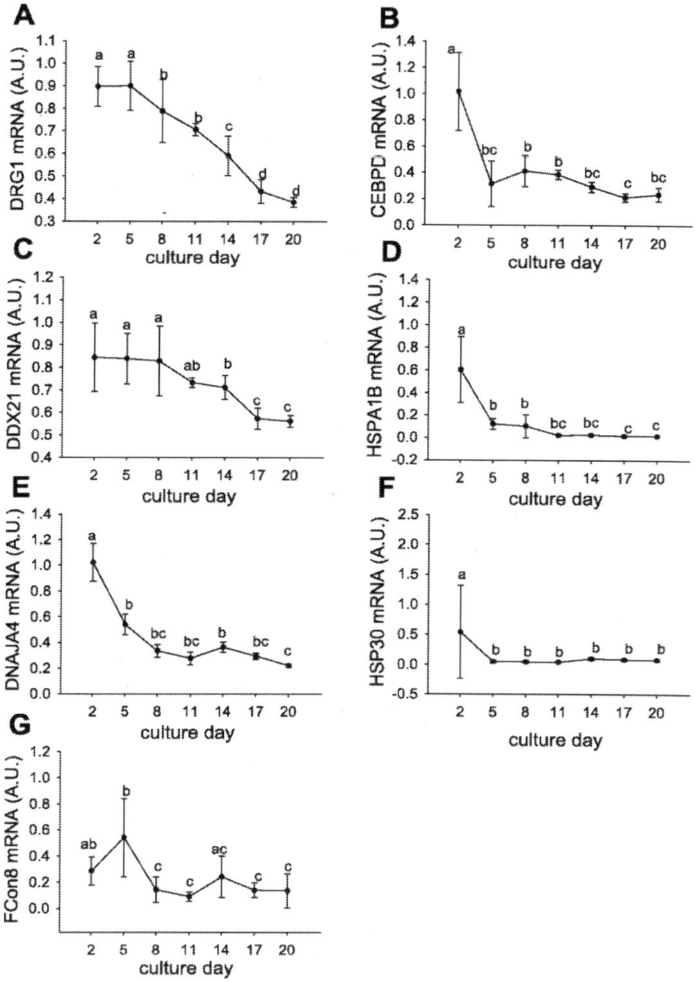

Figure: Expression of candidates in primary myogenic cultures with peak expression at 2–5 d. Expression profiles for *DRG1* (*A*), *CEBPD* (*B*), *DDX21* (*C*), *HSPA1B* (*D*), *DNAJA4* (*E*), *HSP30* (*F*), and *FCon8* (*G*) in Atlantic salmon myogenic cell culture at 2, 5, 8, 11, 14, 17, and 20 d. Values represent means ± SD, 6 samples per point. Significant differences between means are indicated by different letters.

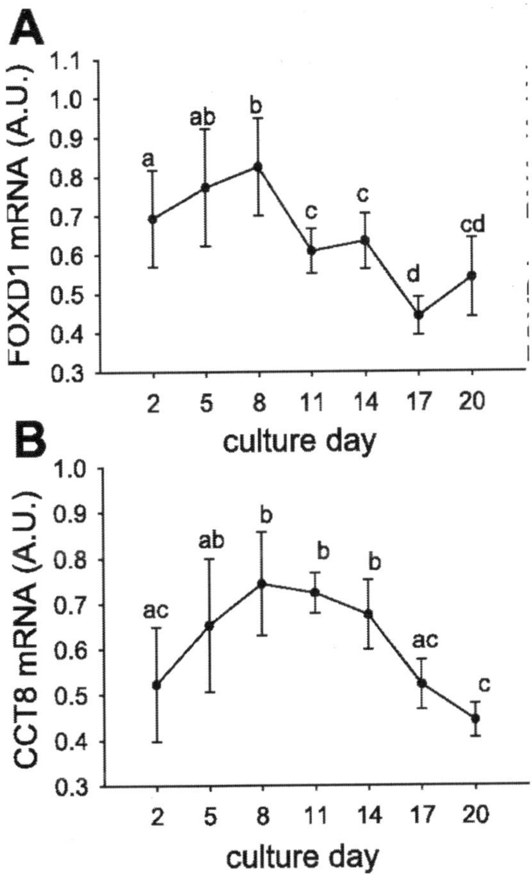

Figure: Expression of candidates in primary myogenic cultures with peak expression at 5 and 8 d. Expression profiles are shown for *FOXD1* (*A*) and *CCT8* (*B*) in Atlantic salmon myogenic cell culture at 2, 5, 8, 11, 14, 17, and 20 d. Values represent means ± SD, 6 samples per point. Significant differences between means are indicated by different letters.

We have identified 23 previously unrecognised genes in fast myotomal muscle of Atlantic salmon that were differentially regulated with growth stimulation in two feeding experiments that involved a distinct population, lifecycle stage, and set of environmental conditions.

Thirteen of these genes were also transcriptionally regulated during the maturation of primary myogenic cell culture and are thus excellent candidates for playing a role in myogenesis. To characterise further the identified genes we examined their expression in the context of other genes with well-characterised roles in myogenesis and growth regulation.

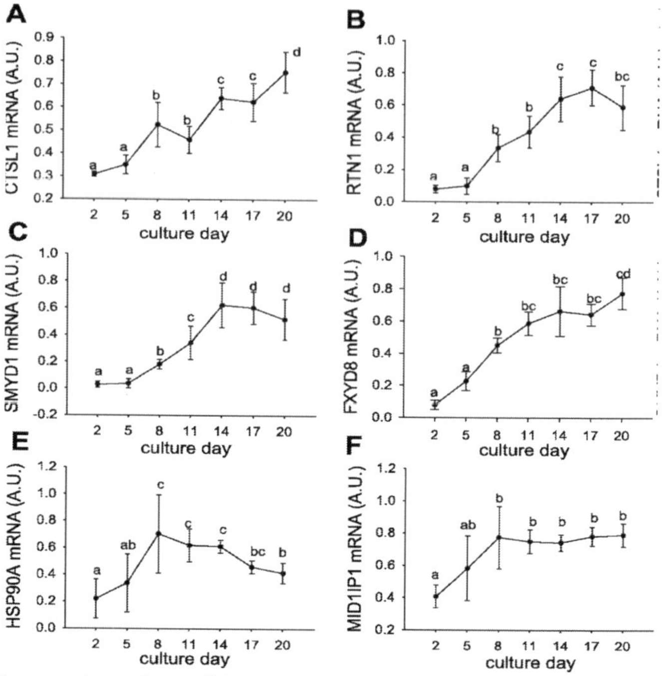

Figure: Expression of candidates in primary myogenic cultures with peak expression coincident with the formation of confluent multinucleated myotubes during late stages of differentiation. *CTSL1* (*A*), *RTN1* (*B*), *SMYD1* (*C*), *FXYD8* (*D*), *HSP90A* (*E*), and *MID1IP1* (*F*). Data points are from cell culture at 2, 5, 8, 11, 14, 17, and 20 d. Values represent means ± SD, 6 samples per point. Significant differences between means are indicated by different letters.

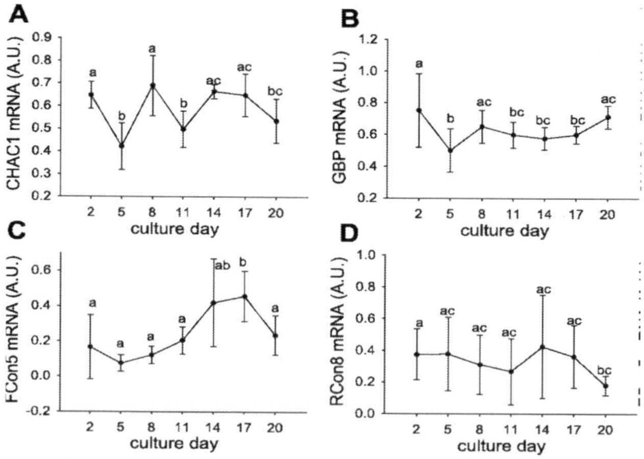

Figure: Expression profiles for genes that formed a subcluster with MRF4. Expression profiles are shown for *CHAC1* (*A*), *GBP* (*B*), *FCon5* (*C*), and *RCon8* (*D*) in Atlantic salmon myogenic cell culture at 2, , 8, 11, 14, 17, and 20 d. Values represent means ± SD, 6 samples per point. Significant differences between means are indicated by different letters.

Genes Associated with Muscle Fibre Differentiation and Positive Growth

The transition from a proliferative to differentiation state requires the coordinated expression of prodifferentiation genes and silencing of genes involved in proliferation, a process that requires alterations in chromatin structure. Two of the genes we identified, *SMYD1* and *RTN1*, may play opposing roles in chromatin modification. Within the tissues examined, *SMYD1* expression was largely restricted to cardiac, slow and fast muscle.

During myotube maturation, *SMYD1* expression increased >30-fold from 2 d to 14 d and remained at elevated levels for 17–20 d when myotubes predominated. Its expression was highly correlated with that of *MEF2A* (R = 0.84, $P < 0.001$) and *IGF2R* (R = 0.78, $P < 0.001$), giving strong evidence for a role in myogenic differentiation.

The SMYD1 protein contains evolutionary conserved SET and MYND domains with histone methyltransferase and deacetylation activity, respectively, indicative of a role in chromatin modification leading to transcriptional suppression. Indeed, acetylation of histones results in decreased levels of transcription, and Gottlieb et al. demonstrated that SMYD1 can function as a transcriptional repressor in mouse skeletal muscle. SMYD1 is expressed in the developing skeletal muscle cells of the somite as well as differentiated muscle fibres in mice and is required for normal skeletal muscle development in zebrafish.

RTN1, a member of the poorly understood reticulon gene family, increased in expression 21-fold during refeeding. Its expression clustered with *IGFBP-4* and was correlated with *IGF-I* (R = 0.82, $P < 0.001$) in vivo and with *MEF2A* (R = 0.84, $P < 0.001$) and *IGF2R* (R = 0.78, $P < 0.001$) in myogenic culture. Correlations with such genes are indicative of a role in myogenic differentiation. Through interaction with SNARE proteins RTN1 is involved in vesicle trafficking and exocytosis, including growth hormone secretion. *RTN1* is also potentially a nuclease and an inhibitor of histone deacetylases, a function that leads to the dissociation of histones from DNA resulting in increased transcription. Thus, whereas SMYD1 may function to repress the expression of genes that inhibit myogenic differentiation, RTN1 may promote the expression of genes required for differentiation. Interestingly, *SMYD1* and *RTN1* expression during myotube maturation was highly correlated (R = 0.97, $P < 0.0001$), which suggests that these genes share similar regulatory elements or are present on the same signalling pathway.

Microtubules play important roles in many myogenic processes including cell morphology and sarcomere assembly and are reorganised to form stable microtubules during differentiation (32). Proteins that stabilise microtubule formation have been shown to be essential for skeletal muscle cell differentiation (77). Mid1 interacting protein (MID1IP1), also known as MIG12, stabilises microtubule formation through its interaction with Mid1 and is implicated in processes requiring microtubule stabilisation (5). We observed that expression of *MID1IP1* increased during myotube maturation, clustered with genes involved in differentiation and was correlated with *myog* (R = 0.76, $P < 0.001$). This, together with its high expression level in slow and fast muscle relative to other tissues, suggests an involvement in the differentiation of myogenic cells.

HSP90A is required for the folding and activation of proteins such as myoD and many proteins involved in signal transduction. Inhibition of HSP90 activity disrupted zebrafish somite development, decreased myogenin expression, inhibited myotube formation, and depleted levels of three protein kinases in C2C12 cells, including AKT. During refeeding, we observed a biphasic response in *HSP90A* expression, with increased expression at 1 d, which then decreased to 0 d values by 5 d and then increased again for 8 d to 21 d. This response likely represents the diverse functions of this protein during the unfolded protein response (UPR). At later time points, the observed increase in expression is likely associated with differentiation processes, as during myogenic cell culture we observed correlated gene expression with *myoD1a* (R = 0.83, $P < 0.001$) and *myog* (R = 0.88, $P < 0.001$).

Genes Associated With Cell Cycle Control and Myoblast Proliferation

Growth and regeneration of muscle requires the proliferation of myogenic progenitor cells from a pool of metabolically quiescent cells in a process that is activated by exercise, muscle damage, and degenerative muscle diseases. There is evidence that fasting checks myogenic progenitors at the G1/S stage of the cell cycle in teleosts (17). Fasting and refeeding stimulates proliferation of fish myogenic cells in vivo and in vitro. Two of the candidate genes had expression profiles characteristic of genes playing a role in the proliferation of myoblasts, or controlling entry into the cell cycle. Transcripts of *FOXD1*, a poorly characterised transcription factor, increased significantly 1 d after feeding, decreased to 0 d values by 8 d and clustered with *myoD1b*.

Interestingly, during myogenic culture its expression was elevated at early time points as cells started to proliferate but then decreased as cells differentiated. Expression during myogenesis clustered with *myoD1b*, *myoD1c* and was highly correlated with *Pax7* (R = 0.68, $P < 0.001$) and *IGF-I* (R = 0.67, $P < 0.001$), all of which are expressed in proliferating myoblasts. This, combined with its high expression largely restricted to fast muscle, suggests a possible role for this transcription factor in myoblast proliferation. Interestingly, in T-cells, FOXD1 regulates FOXJ1, to suppress the NF-ê â-pathway, which is implicated in muscle wasting.

DRG1 expression was also increased at early time points during refeeding, with levels returning to 0 d values by 21 d; however, in myogenic culture, expression started to decrease after 5 d and was correlated with *IGF-II* (R = 0.72, $P < 0.001$). This expression pattern suggests that *DRG1* is also expressed in proliferating cells, which increase in number during refeeding and decrease in number as the cell culture progressed and differentiated cells predominated. *DRG1* has been implicated in regulation of the cell cycle and may be important in growth and differentiation of cells.

As well as genes expressed during the proliferative and differentiation phases of myogenesis, we identified the leucine zipper family member *CEBPD*, which appears to be specifically expressed in quiescent cells. *CEBPD* expression is highly upregulated during G0 phase in growth-arrested mammary epithelial cells and in human limbal stem cells, where it regulates the cell cycle and promotes self renewal of holo-clone forming cells. Although *CEBPD* is strongly upregulated during the differentiation of adipocytes in mammals, in salmon it was 21-fold downregulated with feeding, whereas *adipoQ*, a marker of adipocyte differentiation was upregulated. *CEBPD* was also downregulated concomitant with myoblast proliferation in vitro. It is therefore likely that *CEBPD* is expressed in quiescent myogenic progenitor cells in teleosts. *CEBPD* is regulated by CREB through binding to CREB elements in the promoter of *CEBPD*, and we found a strong correlation between *CREBA* and *CEBPD* expression (R = 0.73, $P < 0.001$).

Genes Associated With Protein Degradation Pathways

A microarray study involving rainbow trout (Oncorhynchus mykiss) revealed 1,000 transcripts that were overexpressed in fast muscle of fish fasted for 30 d and then downregulated after 4 d of feeding, with a preponderance of transcripts involved in ubiquitin-proteasomal and lysosomal protein degradation pathways and fatty acid oxidation (61).

Similar changes were found in a comparison of gravid and sterile rainbow trout (69), and refeeding of fasted salmon also results in the downregulation of genes associated with proteasomal protein degradation pathways such as the E3 ubiquitin ligases *MuRF1* and *MAFbx* (14). The present study identified genes that were strongly downregulated in refed fish and were highly correlated with the expression of *MuRF1* and *MAFbx*. *CTSL1*, a cysteine protease involved in the lysosomal degradation of proteins, was strongly downregulated as soon as 1 d of feeding and expression highly correlated with the E3 ubiquitin ligases MuRF1 and MAFbx (R = 0.89, $P < 0.001$ and R = 0.93, $P < 0.001$, respectively). This expression suggests that during fasting, protein degradation occurs via both proteasomal and lysosomal systems.

RCon2 has homology to a novel protein with similarities to the vertebrate ankyrin repeat-containing protein with a suppressor of cytokine signalling box-2 (ASB2) gene. ASB2 interacts with Elongin BC complex and assembles with Cullin 5/Rbx1, forming an E3 ubiquitin complex. In myeloid leukemia cells, ASB2 responds to retinoic acid, inhibits growth, and promotes differentiation. ASB2 targets filamin B for proteasomal degradation, and during differentiation of C2C12 myoblasts its knockdown delays myoblast fusion and the expression of muscle contractile proteins. We observed that the high expression of *ASB2* in fasted fish was strongly downregulated in response to feeding and was restricted to skeletal muscle. However, expression during myogenesis was below the limit of detection by qPCR, which, considering the high expression we observed in fish muscle (Cq values 22–28 cycles), suggests that this protein is expressed specifically in quiescent cells or expressed in another cell type other than myoblasts, similar to the restricted expression of ASB6 in adipocytes. The expression pattern we observe suggests that *S. salar ASB2* is not involved in muscle differentiation but may play a role in growth inhibition, as has been demonstrated for other ASB family members.

Interestingly, *RCon3* expression was highly correlated with *ASB2* (R = 0.91, $P < 0.001$), was also restricted to skeletal muscle, and was expressed below the limits of detection during myogenesis. BLAST searches using this contig found no homology to known genes; however, further in silico analysis of this sequence revealed significant homology (91% identity) with *ASB2*, which, combined with its similar expression pattern, suggests that this sequence may be a paralog of *ASB2* or a related family member.

Protein Chaperones are Activated in Response to Feeding

During refeeding of fasted fish we observe increased expression of genes with homology to several heat shock proteins including *DNAJA4, HSPA1B*, and *HSP90A* and a gene with homology to *CHAC1*. The contigs corresponding to these genes were some of the largest in the EST collection with 31 and 100 ESTs corresponding to *CHAC1* and *HSP90A*, respectively. The expression of these genes may be indicative of activation of the unfolded protein response (UPR) pathway, a stress response mechanism activated when endoplasmic reticulum (ER) stress occurs, due to the accumulation of unfolded proteins in the ER. Accumulation of unfolded proteins can occur when the amounts of newly synthesized proteins exceeds that of the protein folding capacity in the ER.

This situation is likely to occur during the transition from a fasted state to that of satiation feeding where increased protein synthesis occurs through activation of the AKT/mTOR pathway. Stimulation of the UPR leads to the activation of several transcription factors, resulting in the expression of *DNAJ, HSPA1B, HSP90*, and *CHAC1*. There are three different UPR response pathways, regulated by PERK, ATF-6, and IRE-1. Activation of the PERK pathway leads to phosphorylation of eIF2á, resulting in reduced translation. The ATF-6 pathway leads to the induction of ER chaperones involved in protein folding, and the IRE-1 pathway increases the production of ER-associated degradation factors involved in the degradation of unfolded proteins. Our observations suggest that two of these pathways are activated during refeeding of fasted fish, with the heat shock proteins characteristic of the ATF-6 pathway, and *Chac1* indicative of PERK pathway activation. Thus, with refeeding, unfolded proteins are cleared from the ER through decreased rates of translation and increased folding capacity, ensuring the quality of the secreted proteins. This response appears to be short lived as levels of *DNAJA4, HSPA1B,* and *CHAC1* were reduced by *day 3*, presumably as levels of the ER chaperones have reached levels sufficient to cope with the increased flux through the ER. These data also highlight the increased production of secreted proteins that occurs in muscle during the refeeding response.

Genes of Unknown Identity

The identity of the majority of the contigs analyzed by qPCR could be assigned through homology based sequence searches; however, several contigs had no homology to known genes including *FCon5, FCon8, FCon14, RCon3*, and *RCon8*. By retrieving salmonid ESTs from dbEST (9) with homology to these unknown sequences, we were able to generate longer contigs giving far greater sequence coverage; however, the resulting contigs still had no homology to known genes and could represent novel genes, pseudogenes, or untranslated regions. Two of these contigs had expression patterns that make them interesting candidates as promyogenic factors. *FCon5* expression was restricted to skeletal muscle and was highest in fast muscle. During refeeding, expression was elevated at all time points, clustered with that of *IGF-I* and *IGFBP-4*, and was positively correlated with that of *MRF4* in myogenic cell culture ($R = 0.68$, $P < 0.001$). *FCon8* had peak expression at 1–5 d that remained elevated at all time points following refeeding, was most abundant in fast muscle, and was transiently increased at 5 d of myogenic culture. Bioinformatic analysis reveals that these contigs are restricted to salmonids and do not contain an open reading frame. Therefore, if they are part of a gene, they are likely to represent untranslated regions.

We have screened nutritionally regulated genes in vivo against their expression pattern in cell culture in an attempt to identify genes specifically involved in muscle growth. However, this approach has limitations since in vivo muscle is composed of many cell types including fibroblasts and adipocytes with known cross talk and involvement in muscle fibre growth (1). It is therefore not possible to exclude roles in muscle growth for the remaining genes without further investigation.

In summary, we have identified several genes that have expression patterns restricted to muscle, increased expression during refeeding, and, based on their expression in myotube maturation, may be involved in proliferation (*FOXD1* and *DRG1*) and differentiation (*SMYD1, RTN1, MID1IP1, HSP90A*, and *FCon5*) phases of myogenesis and are thus excellent candidates as promyogenic genes. We have also identified genes that could be considered inhibitory to myogenesis, or expressed in quiescent cells (*CEBPD, ASB2, RCon3, and HSP30*). Most of these genes have human orthologs; however, we also identified genes that may be salmonid specific (*FCon5* and *FCon8*) and could be important regulators of growth in Atlantic salmon, a species of indeterminate growth. Future analysis of these candidate genes using gene knockdown experiments in zebrafish could shed further light on their possible roles in myogenesis.

Gonadotropin-Releasing Hormone

Gonadotropin-releasing hormone (GnRH), also known as Luteinizing-hormone-releasing hormone (LHRH) and luliberin, is a trophic peptide hormone responsible for the release of follicle-stimulating hormone (FSH) and luteinizing hormone (LH) from the anterior pituitary. GnRH is synthesized and released from neurons within the hypothalamus. The peptide belongs to gonadotropin-releasing hormone family.

Production

The gene, *GNRH1*, for the GnRH precursor is located on chromosome 8. In mammals, the linear decapeptide end-product is synthesized from a 92-amino acid preprohormone in the preoptic anterior hypothalamus.

Structure

The identity of GnRH was clarified by the 1977 Nobel Laureates Roger Guillemin and Andrew V. Schally:

pyroGlu-His-Trp-Ser-Tyr-Gly-Leu-Arg-Pro-Gly-NH2

GNRH as a Neurohormone

GnRH is considered a neurohormone, a hormone produced in a specific neural cell and released at its neural terminal. A key area for production of GNRH is the preoptic area of the hypothalamus, which contains most of the GnRH-secreting neurons. GnRH neurons originate in the nose and migrate into the brain, where they are scattered throughout the medial septum and hypothalamus and connected by very long >1-millimetre-long dendrites. These bundle together so they receive shared synaptic input, a process that allows them to synchronize their GnRH release.

GnRH is secreted in the hypophysial portal bloodstream at the median eminence. The portal blood carries the GnRH to the pituitary gland, which contains the gonadotrope cells, where GnRH activates its own receptor, gonadotropin-releasing hormone receptor (GnRHR), a seven-transmembrane G-protein-coupled receptor that stimulates the beta isoform of Phosphoinositide phospholipase C, which goes on to mobilise calcium and protein kinase C. This results in the activation of proteins involved in the synthesis and secretion of the gonadotropins LH and FSH. GnRH is degraded by proteolysis within a few minutes.

Control of Fish and LH

At the pituitary, GnRH stimulates the synthesis and secretion of the gonadotropins, follicle-stimulating hormone (FSH), and luteinizing hormone (LH). These processes are controlled by the size and frequency of GnRH pulses, as well as by feedback from androgens and estrogens. Low-frequency GnRH pulses lead to FSH release, whereas high-frequency GnRH pulses stimulate LH release.

There are differences in GnRH secretion between females and males. In males, GnRH is secreted in pulses at a constant frequency, but, in females, the frequency of the pulses varies during the menstrual cycle, and there is a large surge of GnRH just before ovulation.

GnRH secretion is pulsatile in all vertebrates, and is necessary for correct reproductive function. Thus, a single hormone, GnRH1, controls a complex process of follicular growth, ovulation, and corpus luteum maintenance in the female, and spermatogenesis in the male.

Activity

GnRH activity is very low during childhood, and is activated at puberty. During the reproductive years, pulse activity is critical for successful reproductive function as controlled by feedback loops. However, once a pregnancy is established, GnRH activity is not required. Pulsatile activity can be disrupted by hypothalamic-pituitary disease, either dysfunction (i.e., hypothalamic suppression) or organic lesions (trauma, tumor). Elevated prolactin levels decrease GnRH activity. In contrast, hyperinsulinemia increases pulse activity leading to disorderly LH and FSH activity, as seen in polycystic ovary syndrome (PCOS). GnRH formation is congenitally absent in Kallmann syndrome.

The GnRH neurons are regulated by many different afferent neurons, using several different transmitters (including norepinephrine, GABA, glutamate). For instance, dopamine appears to stimulate LH release (through GnRH) in estrogen-progesterone-primed females; dopamine may inhibit LH release in ovariectomized females. Kisspeptin appears to be an important regulator of GnRH release. GnRH release can also be regulated by estrogen. It has been reported that there are kisspeptin-producing neurons that also express estrogen receptor alpha.

GNRH in Other Organs

GnRH is found in organs outside of the hypothalamus and pituitary, and its role in other life processes is poorly understood. For instance, there is likely to be a role for GnRH1 in the placenta and in the gonads. GnRH and GnRH receptors are also found in cancers of the breast, ovary, prostate, and endometrium.

Medication

GnRH is available as gonadorelin hydrochloride(Factrel®)and gonadorelin diacetate tetrahydrate(Cystorelin®) for injectable use. Studies have described it being used via an infusion pump system to induce ovulation in patients with hypothalamic hypogonadism. It is also used in veterinary medicine as a treatment for cattle with cystic ovarian disease.

Its analogue Leuprolide is used for continuous infusion, to treat Breast carcinoma, endometriosis, prostate carcinoma, and precocious puberty.

Agonists and Antagonists

While GnRH has been synthesized and become available, its short half-life requires infusion pumps for its clinical use. Modifications of the decapeptide structure of GnRH have led to GnRH1 analog medications that either stimulate (GnRH1 agonists) or suppress (GnRH antagonists) the gonadotropins. It is important to note that, through downregulation, agonists are also able to exert a prolonged suppression effect.

Osmoregulation

Osmoregulation is the active regulation of the osmotic pressure of an organism's fluids to maintain the homeostasis of the organism's water content; that is it keeps the organism's fluids from becoming too diluted or too concentrated. Osmotic pressure is a measure of the tendency of water to move into one solution from another by osmosis. The higher the osmotic pressure of a solution the more water wants to move into the solution. Pressure must be exerted on the hypertonic side of a selectively permeable membrane to prevent diffusion of water by osmosis from the side containing pure water.

Organisms in both aquatic and terrestrial environments must maintain the right concentration of solutes and amount of water in their body fluids; this involves excretion (getting rid of metabolic wastes and other substances such as hormones that would be toxic if allowed to accumulate in the blood) via organs such as the skin and the kidneys; keeping the amount of water and dissolved solutes in balance is referred to as osmoregulation.

Regulators and Conformers

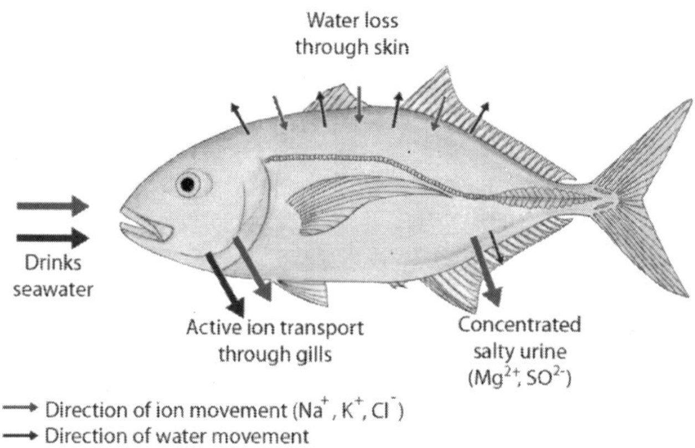

Figure: Movement of water and ions in saltwater fish

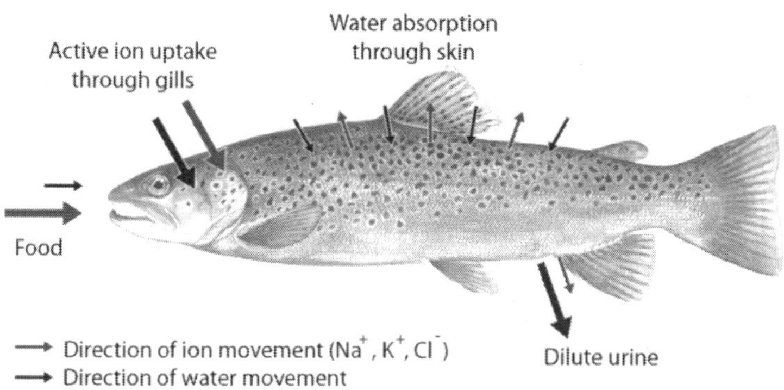

Figure: Movement of water and ions in freshwater fish

Two major types of osmoregulation are osmoconformers and osmoregulators. Osmoconformers match their body osmolarity to their environment. It can be either active or passive. Most marine invertebrates are osmoconformers, although their ionic composition may be different from that of seawater.

Osmoregulators tightly regulate their body osmolarity, which always stays constant, and are more common in the animal kingdom. Osmoregulators actively control salt concentrations despite the salt concentrations in the environment. An example is freshwater fish. The gills actively uptake salt from the environment by the use of mitochondria-rich cells. Water will diffuse into the fish, so it excretes a very hypotonic (dilute) urine to expel all the excess water. A marine fish has an internal osmotic concentration lower than that of the surrounding seawater, so it tends to lose water and gain salt. It actively excretes salt out from the gills. Most fish are stenohaline, which means they are restricted to either salt or fresh water and cannot survive in water with a different salt concentration than they are adapted to. However, some fish show a tremendous ability to effectively osmoregulate across a broad range of salinities; fish with this ability are known as euryhaline species, e.g. Salmon.

Some marine fish have adopted a different but efficient mechanism to conserve water, i.e. osmoregulation. They retain urea in their blood in relatively higher concentration; this helps them to retain a nearly orgasmic environment with higher solute but lower salt concentration. However, urea is damaging to living tissue, so, to cope with this problem some fish retain *trimethylamine oxide*, this provides a better solution to urea's toxicity.

Osmoregulation in Plants

Whilst there are no specific osmoregulatory organs in higher plants the stomata are important in regulating water loss through evapotranspiration and on the cellular level the vacuole is crucial in regulating the concentration of solutes in the cytoplasm. Strong winds, low humidity and high temperatures all increase evapotranspiration from leaves. Abscisic acid is an important hormone in helping plants to conserve water - it causes stomata to close and stimulates root growth so that more water can be absorbed.

Plants share with animals the problems of obtaining water but unlike in animals the loss of water in plants is crucial to create a driving force to move nutrients from the soil to tissues. Certain plants have evolved methods of water conservation. Xerophytes are plants that can survive in dry habitats, such as deserts, and are able to withstand prolonged periods of water shortage. Succulent plants such as the cacti store water in the vacuoles of large parenchyma tissues. Other plants have leaf modifications to reduce water loss, such as needle-shaped leaves, sunken stomata, and thick, waxy cuticles as in the pine.

The sand-dune marram grass has rolled leaves with stomata on the inner surface. Hydrophytes are plants in water habitats. They mostly grow in water or in wet or damp places. In these plants the water absorption occur through the whole surface of the plant, e.g. the water lily.

Osmoregulation in Protists and Animals

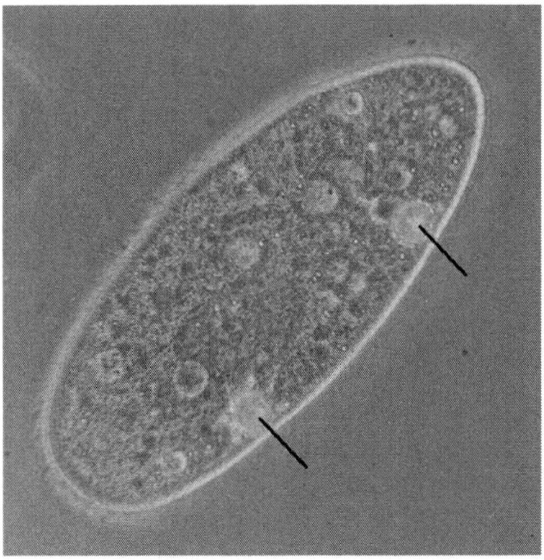

Figure: Protist Paramecium aurelia with contractile vacuoles.

Amoeba make use of contractile vacuoles to collect excretory waste, such as ammonia, from the intracellular fluid by both diffusion and active transport. As osmotic action pushes water from the environment into the cytoplasm, the vacuole moves to the surface and disposes the contents into the environment.

Kidneys play a very large role in human osmoregulation, regulating the amount of water in urine waste. With the help of hormones such as antidiuretic hormone, aldosterone, and angiotensin II, the human body can increase the permeability of the collecting ducts in the kidney to reabsorb water and prevent it from being excreted.

A major way animals have evolved to osmoregulate is by controlling the amount of water excreted through the excretory system.

Vertebrate Excretory Systems

Waste Products of the Nitrogen Metabolism: Ammonia is a toxic by-product of protein metabolism and is generally converted to less toxic substances after it is produced then excreted; mammals convert ammonia to urea, whereas birds and reptiles form uric acid to be excreted with other wastes via their cloacas.

Achieving Osmoregulation in Vertebrates

Four processes occur:

- filtration - fluid portion of blood (plasma) is filtered from a nephron (functional unit of vertebrate kidney) structure known as the glomerulus into Bowman's capsule or glomerular capsule (in the kidney's cortex) and flows down the proximal convoluted tubule to a "u-turn" called the Loop of Henle (loop of the nephron) in the medulla portion of the kidney.
- reabsorption - most of the viscous glomerular filtrate is returned to blood vessels that surround the convoluted tubules.
- secretion - the remaining fluid becomes urine, which travels down collecting ducts to the medullary region of the kidney.
- excretion - the urine (in mammals) is stored in the urinary bladder and exits via the urethra; in other vertebrates, the urine mixes with other wastes in the cloaca before leaving the body (frogs also have a urinary bladder).

Tissue Hydration

Tissue hydration is the process of absorbing and retaining water in biological tissues.

Plants

Land plants maintain adequate tissue hydration by means of an outer waterproof layer. In soft or green tissues, this is usually a waxy cuticle over the outer epidermis. In older, woody tissues, waterproofing chemicals are present in the secondary cell wall that limit or inhibit the flow of water. Vascular plants also possess an internal vascular system that distributes fluid throughout the plant.

Some xerophytes, such as cacti and other desert plants, have mucilage in their tissues. This is a sticky substance that holds water within the plant, reducing the rate of dehydration. Some seeds and spores remain dormant until adequate moisture is present, at which time the seed or spore begins to germinate.

Animals

Animals maintain adequate tissue hydration by means of (1) an outer skin, shell, or cuticle; (2) a fluid-filled coelom cavity; and (3) a circulatory system.

In humans, a significant drop in tissue hydration can lead to the medical condition of dehydration. This may result from loss of water itself, loss of electrolytes, or a loss of blood plasma.

Some invertebrates are able to survive extreme desiccation of their tissues by entering a state of cryptobiosis.

Salt Gland

Figure: Sea turtles excrete salts through tear ducts. "Crying" is visible when out of water.

The salt gland is an organ for excreting excess salts. It is found in elasmobranchs, seabirds, and some reptiles. In sharks, salt glands are found in the rectum, but in birds and reptiles, they are found in or on the skull, in the area of the eyes, nostrils or mouth. In crocodiles, the salt is excreted through the tongue. Such glands work by active transport via sodium-potassium pump that moves salt from the blood into the gland, where it can be excreted as a concentrated solution. Salt glands function to maintain salt balance and allow marine vertebrates to drink seawater.

The aviary salt gland is coupled with a counter-current exchange mechanism that preserves the high concentration of salt in the blood plasma so that most of the salt leaving the gland is returned to the incoming blood plasma, allowing the concentration buildup within the gland by active transport.

The need for salt excretion in reptiles (such as marine iguanas and sea turtles) and birds (such as petrels and albatrosses) stems from the fact that their kidneys are much less efficient than those of mammals. Unlike the skin of amphibians, reptile and bird skin is impermeable to salt, meaning that the transition to a tougher skin meant a loss in salt-releasing ability. The evolution of a salt gland would have allowed early reptiles and birds to eat aquatic plants and animals, who have high salt concentrations. This does not, however, explain the evolution of the gland in the elasmobranchs, suggesting convergent evolution.

Some theories suggest that mammalian tear ducts and sweat glands may be evolutionarily related to salt glands. Human tears are high in potassium, lending support to this theory; however, most phylogeneticists disagree with this idea.

Osmoregulation in Freshwater Fish

Fish live their lives completely surrounded by water on all sides. It surrounds them externally in their habitat and also comprises much of their body mass. Fish must therefore strike some sort of balance between these two separate, and very different, water environments which determine so much about their health. This balance is met through the processes of osmosis and osmoregulation.

To explain osmosis, we'll first discuss the concepts of diffusion and equilibrium. All molecules have energy associated with them known as kinetic energy. This energy of motion allows for the random movement of molecules throughout the space which they are in. Diffusion is simply the tendency of molecules to passively move from an area where they are more highly concentrated to an area where they are at a lower concentration - as a result of random movement. Note here, that it is the molecules as a *whole* which move from higher to lower areas of concentration.

The movement of *individual* molecules is random and while some may move down the concentration gradient (from higher to lower concentration), others will move up the concentration gradient. There will be a *net* movement of molecules down the concentration gradient until a point of equilibrium is reached. At this point, the molecules are moving at equal rates up and down the concentration gradient and there is no further net movement of the particles. Diffusion and equilibrium can be illustrated by placing a drop of methylene blue in a glass of water. One will observe that after some time has passed, this dye has dispersed throughout the water until an even colour can be observed in all areas. The dye molecules diffused through the water randomly and reached a point of equilibrium such that there is no further net colour change in the glass. Keep in mind that although there are equal numbers of dye molecules in all areas of the glass, individual molecules aren't static and don't remain in the same place. It is merely their concentration which is maintained. In a different scenario, one can imagine being in a large room where someone lights a cigarette. At first only the people nearest to the cigarette will smell the smoke. However, after some time, everyone in the room will smell it. When the cigarette is no longer burning, the scent will be evenly detectable throughout the room as the smoke molecules diffuse and reach equilibrium.

Osmosis is a concept similar to diffusion which involves the passive movement of water through a membrane which is permeable to the water, but not to the solutes dissolved in the water. With osmosis, the water moves from an area of lower solute concentration to an area of higher solute concentration. While this may initially seem to be the opposite of diffusion, one must consider that the water is still moving from where the water is most abundant to where it is least abundant.

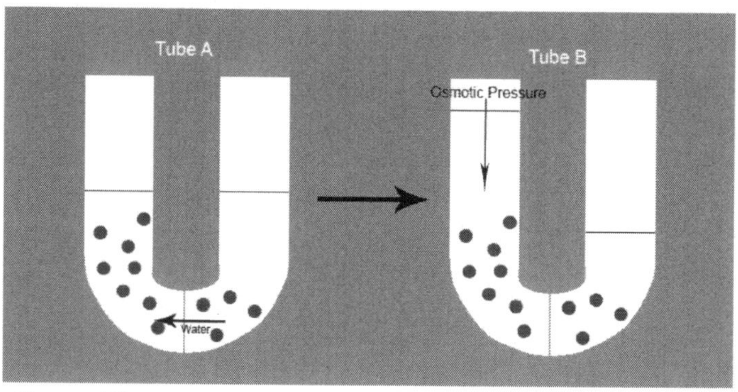

The figure illustrates two U-shaped tubes, each of which contain two different salt solutions separated by a semi-permeable membrane which allows only water to pass through it. In Tube A, the left side contains a hypertonic solution - having a higher concentration of solutes, and the right side contains a hypotonic solution, having a lower concentration of solutes. Isotonic is a term for solutions of equal solute concentration. Water moves from the hypotonic side to the hypertonic side and in doing so, increases the salt concentration on the right side of the tube and decreases the salt concentration on the left side of the tube until the two concentrations are equal. Tube B illustrates equilibrium. At equilibrium water is moving though the membrane in both directions at an equal rate. The pressure needed to stop the water level from rising on the left side of the tube is called the "osmotic pressure". Water always has a net movement toward the solution of higher osmotic pressure. At equilibrium, both sides of the tube have equal osmotic pressures.

The biological importance of hypertonic, hypotonic, and isotonic solutions and their effects on living organisms can be observed by subjecting a single cell to each of them, when the cell has no means of adapting to each new environment. A cell placed into a hypotonic solution will have water rush inside of it (where solutes are more concentrated) and cause the cell to burst. A cell placed into a hypertonic solution will experience dehydration as water leaves the cell for the surrounding environment. A cell placed into an isotonic solution will experience water entering the cell at the same rate at which it leaves, and so will be at equilibrium and appear normal.

Fish do not always find themselves in isotonic environments. Thus, their body cells must have a means by which to adapt to changing salt concentrations in their bodies and environments. Osmoregulation controls this balance of water/salt concentrations. Freshwater fish are hypertonic to their water environment and therefore, water is continually diffusing into the fish through the gill membranes into the blood. The gills are also permeable to respiratory gases, ammonia waste products, and ions.

Therefore, while water moves in towards the higher osmotic pressure of the blood, sodium and chloride ions also diffuse out of the fish, moving down their concentration gradients to the external environment. Freshwater fish must expend energy to regulate this ion loss and fluid uptake. Marine fish experience the opposite situation as their bodies are hypotonic to their saltwater environment.

The continual uptake of water in freshwater species is regulated by the kidneys which continually produce large amounts of dilute urine. Despite the importance of healthy kidneys to help counteract the problem of taking on water, some salts are also lost in the large amounts of urine as well as through the membrane of the gills.

Fortunately, the gills are also a site of ion uptake. Special cells in gill lamellae contain sodium and chloride "pumps". These pumps are special enzymes that use energy to move the ions up their concentration gradient (remember that moving down a concentration gradient is spontaneous, as in diffusion, and requires no input of energy) to maintain their higher concentration in the body.

Thus, osmoregulation is a process that requires the expenditure of much energy on the part of the fish—even when they appear to be inactive. This constant expenditure of energy to maintain an osmotic balance is a reason why proper nutrition and low stress levels are important for healthy fish. Damage to the kidneys through bacterial infection or other means is often deadly as these organs extract the large amounts of water which continually diffuse into the fish's body. Fish with ascites or bloat are often suffering from kidney damage which is, unfortunately, irreversible. Saving a fish which appears bloated is highly unlikely, if the swelling is due to kidney dysfunction as they have lost the ability of osmoregulation.

Angelfish and discus are examples of fish that have very strong and efficient osmoregulatory systems. They do well naturally in their native environment where the osmotic pressure is great due to the extremely soft water of the Amazon river basin. However, when under stress, these systems can be impaired. This is why it is advised to add some salt to the water of fish under stress. The salt reduces the osmotic pressure. The stress could come from transportation, diseases, trauma or a number of other problems. Keep in mind that salt is only good for short term success. The stress must be removed to prevent further problems.

A sudden change in osmotic pressure can put great stress on the osmoregulatory system of a fish. This is of great concern when shipping fish to locations with water different from what they're adapted to. The fish arrives under great stress and is not able to regulate any osmotic pressure differences easily. This is one reason why acclimation should be slow. It also explains why treating diseased fish must be done carefully. When putting them in a salt bath, the concentration of salt should be increased gradually.

It is important to understand osmosis and how it affects our fish. It is a vital component for their well-being. Your ability to control problems, safely ship and receive fish and treat diseases will be enhanced if you pay attention to this aspect of their lives.

Halotolerance

Halotolerance is the adaptation of living organisms to conditions of high salinity. Halotolerant species tend to live in areas such as hypersaline lakes, coastal dunes, saline deserts, salt marshes, and inland salt seas and springs. Halophiles are a group of bacteria that live in highly saline environments, and indeed in many cases require the salinity to survive. Halophytes are salt-tolerant higher plants.

Applications

Fields of scientific research relevant to halotolerance include biochemistry, molecular biology, cell biology, physiology, ecology, and genetics.

An understanding of halotolerance can be applicable to areas such as arid-zone agriculture, xeriscaping, aquaculture (of fish or algae), bioproduction of desirable compounds (such as phycobiliproteins or carotenoids) using seawater to support growth, or remediation of salt-affected soils. In addition, many environmental stressors involve or induce osmotic changes, so knowledge gained about halotolerance can also be relevant to understanding tolerance to extremes in moisture or temperature.

Goals of studying halotolerance include increasing the agricultural productivity of lands affected by soil salination or where only saline water is available. Conventional agricultural species could be made more halotolerant by gene transfer from naturally halotolerant species (by conventional breeding or genetic engineering) or by applying treatments developed from an understanding of the mechanisms of halotolcrancc. In addition, naturally halotolerant plants or microorganisms could be developed into useful agricultural crops or fermentation organisms.

Cellular Functions

Tolerance of high salt conditions can be obtained through several routes. High levels of salt entering the plant can trigger ionic imbalances which cause complications in respiration and photosynthesis, leading to reduced rates of growth, injury and death in severe cases. To be considered tolerant of saline conditions, the protoplast must show methods of balancing the toxic and osmotic effects of the increased salt concentrations. Halophytic vascular plants can survive on soils with salt concentrations around 6%, or up to 20% in extreme cases.

Tolerance of such conditions is reached through the use of stress proteins and compatible cytoplasm osmotic solutes.

To exist in such conditions, halophytes tend to be subject to the uptake of high levels of salt into their cells, and this is often required to maintain an osmotic potential lower than that of the soil to ensure water uptake. High salt concentrations within the cell can be damaging to sensitive organelles such as the chloroplast, so sequestration of salt is seen. Under this action, salt is stored within the vacuole to protect such delicate areas.

If high salt concentrations are seen within the vacuole, a high concentration gradient will be established between the vacuole and the cytoplasm, leading to high levels of energy investment to maintain this state. Therefore, the accumulation of compatible cytoplasmic osmotic solutes can be seen to prevent this situation from occurring. Amino Acids such as proline accumulate in halophytic Brassica species, quaternary ammonium bases such as Glycine Betaine and sugars have been shown to act in this role within halophytic members of Chenopodiaceae and members of Asteraceae show the build up of cyclites and soluble sugars. The build up of these compounds allow for the balancing of the osmotic effect while preventing the establishment of toxic concentrations of salt or requiring the maintenance of high concentration gradients.

Bacterial Halotolerance

The extent of halotolerance varies widely amongst different species of bacteria. A number of cyanobacteria are halotolerant; an example location of occurrence for such cyanobacteria is in the Makgadikgadi Pans, a large hypersaline lake in Botswana.

Stress and Fish Health

Stress is present in the lives of all living things and is the force that brings about physical change and adjustment. Small amounts of stress can be harmless or even beneficial, but high levels of stress or prolonged periods of stress can create severe health problems. Many people are aware of stress in their own lives and can name many of the causes as well as possible treatments. However, the stress that affects fish is different and much more widespread. The nature of keeping aquatic species in confined environments generates many stresses that are unique to aquarium fish. To be successful in keeping healthy aquarium fish, you need to know what causes stress in fish as well as how to prevent it. Elevated stress levels are at the root of most health problems in fish and this article will help identify the causes of stress and give recommendations for treatment.

What is Stress?

Stress is any condition that causes physical or mental discomfort that results in the release of stress-related hormones or results in specific physiological responses. For example, stressful events will cause an increase in heart rate, blood pressure, increased blood sugar, and the release of cortisol. Stress can be physical, psychological, or environmental. Stress can either be short and sudden, or long and chronic. Mild, short-term stress has few serious health effects, but long-term stress or severe, short-term stress contribute to many of the illnesses and deaths in aquarium fish.

The Effect Stress Has on a Fish's Health

Short-term stress will cause an increase in heart rate, blood pressure, and respiration. The fish is reacting much as we do with the fight or flight mode. The fish can only maintain these altered states for a short period of time and then they will adapt or the stress will become chronic. Stress is accompanied by the release of the hormone cortisol, which is responsible for many of the negative health effects associated with stress. In addition to having a negative effect on growth, reproduction, and digestion, chronic stress will also lower the ability of the immune system to respond effectively and fully. This lowered immune response is what allows parasites, bacteria, and fungi to infect a stressed fish and cause disease and death.

The Causes of Stress

There are dozens of potential stresses to fish, but some of the more common causes are:

- Elevated ammonia
- Elevated nitrate
- Improper pH level
- Fluctuations in temperature
- Improper salinity
- Low oxygen levels
- Harassment from other fish
- Lack of hiding places
- Inadequate tank size
- Overstocking of tank
- Medications and water treatments
- Improper nutrition
- Disturbance of the tank
- Harvesting and shipping of fish
- Lack of enough fish to provide schooling

Elevated ammonia, nitrite, and nitrate all create deterioration in fish health due to stress. High levels can cause severe stress, whereas slightly elevated levels can contribute to chronic stress.

pH levels that change abruptly cause acute stress and continually elevated or lowered pH levels can cause chronic stress. Many fish adapt to long-term changes, but there are limits. pH changes of more than 1.5 points below or above recommended levels are going to have a negative effect over time and should never be considered acceptable.

Temperature fluctuations are a much underappreciated stressor of fish. Most tropical freshwater and marine fish do not tolerate temperature changes very well. Many tanks that are not set up properly will have over the recommended maximum of one degree of temperature fluctuation in a 24-hour period due to room temperature, lights, and equipment. The daily fluctuations will create chronic stress as will having a too low or too high of a temperature in the tank for the species of fish present.

Wild fish live within very specific salinity levels (levels of salt in the water). Their bodies work hard to maintain the osmotic gradient between themselves and their environment. If their environmental salinity is not specific to their needs and is not held at a steady level, they have to work harder to maintain their osmotic gradient, which generates chronic stress.

Oxygen levels that are below recommended levels can cause fish to 'breathe' faster than optimum and this can result in chronic stress. Obviously, very low oxygen levels can lead to severe short-term stress and death.

Harassment from other fish and lack of hiding spaces go hand in hand. There should be two suitable hiding spaces for every fish in the tank, otherwise, there are going to be fish that are stressed and bullied. Remember that unlike their environment in the wild, these fish are confined and cannot get away from aggressors. Aggression is a very real problem in many tanks that leads to many injuries, infections, and death.

Overstocking of the tank is a common problem that contributes to almost all of the stresses in the above list, from water pollution to oxygen depletion to harassment. Do not overstock your tanks. If you want to stress your fish, put too many in the tank and it will happen every time.

If you add something to the water to treat a disease or water condition, be aware that it can be stressful to your fish. Try to avoid treating the water if at all possible and always use a quarantine or treatment tank. Copper is an excellent treatment for ich or velvet, but it can be toxic and stressful to fish. Of course, using it is much better than letting a fish die from velvet, but it should never be used in a tank with healthy fish.

Improper nutrition is also a commonly overlooked stressor of fish. Many fish can live on minimal nutrition with old or stale flake foods, but this poor nutrition is a chronic stress. A variety of well-preserved dry foods as well as freeze-dried, fresh, and frozen foods specifically designed for individual species are necessary to prevent chronic nutritional stress.

Disturbing the tank through banging on the glass, constantly netting fish, or rearranging décor stresses fish and should be kept to the necessary minimum.

There is probably nothing that stresses fish more than bringing them from the wild or an aquaculture pond through the wholesaler to your home. In just a few days, the fish will be captured, held, packaged, shipped, sorted, handled, packaged again, and so on through the collector, exporter, importer, wholesaler, and retailer to your tank. Throughout this process they may be exposed to drastic changes in temperature, ammonia, pH, salinity, diet, medications etc. They often do not eat and arrive at your tank completely stressed.

If they are not handled very carefully and are not placed in an optimum environment, their stress is going to continue and they will get sick and die. The reason that this stress is listed last is not because it is the least important, but because it is the most significant.

You need to understand all the stress factors and how to eliminate them because these fish that arrive at your tank have been exposed to all of the listed stresses. The unfortunate truth is that the majority of fish mortalities occur at or near the time of entering a new tank and only through an appreciation of stress and its effect on fish can this problem be prevented.

How you Eliminate Stress

While it is impossible to eliminate all stress, we fortunately have the ability to limit or prevent many of the causes. Acute stress is more obvious and needs to be addressed very quickly. Chronic stress is often not visible. It can take weeks and months to develop. Your fish may appear to be doing fine, until one day one gets sick and dies, and then a few weeks later another one does and so on. If you have fish that are getting sick and dying, there is probably a source of stress on them that needs to be identified and remedied.

The other big source of stress is bringing new fish into your aquarium. Buy only from the most reputable sources that move the fish in the most careful and humane manner. Acclimate your new fish properly, use a quarantine tank, and make sure your new fish are fed appropriately. Spend extra time on the new fish and be as careful as you can. Taking a little time here can make all the difference in breaking the chronic stress cycle and keeping these fish healthy and disease free.

Stress is one of the most critical factors in fish health. Only by understanding the effects that stress have on fish, as well as being able to identify and prevent common stresses, can we eliminate this problem. As aquarists, we need to be responsible for the health and welfare of all of our fish. Provide the highest quality water, nutrition, and suitable tank environment. Introduce new fish carefully and always use a quarantine or treatment tank when necessary. If we work hard to reduce the stress in our fish, we can virtually eliminate disease and health problems in our aquarium.

The Role of Stress in Fish Disease

By R.W. Rottmann, R. Francis-Floyd1, and R. Durborow. Published by the Southern Regional Agricultural Centre and the Texas Aquaculture Extension Service. Physiological stress and physical injury are the primary contributing factors of fish disease and mortality in aquaculture. Stress is defined as physical or chemical factors that cause bodily reactions that may contribute to disease and death.

Stress

Many potential fish disease pathogens are continually present in the water, soil, air, or fish. In nature fish are often resistant to these pathogens, and they are able to seek the best living conditions available. Food fish reared under commercial aquaculture conditions are confined to the production unit and are weakened by stress conditions including:

- increased fish density and poor water quality (i.e., low dissolved oxygen, undesirable temperature or pH, increased levels of carbon dioxide, ammonia, nitrite, hydrogen sulfide, organic matter in the water);
- injury during handling (i.e., capture, sorting, shipping);
- inadequate nutrition; and
- poor sanitation.

These conditions can result in decreased resistance by the fish, resulting in the spread of disease and parasite infestation. Stress and injury initially trigger an alarm reaction (fight or flight response), which results in a series of changes within the fish. A blood sugar increase occurs in response to hormone secretion from the adrenal gland as liver glycogen is metabolized. This produces a burst of energy which prepares the animal for an emergency situation. In addition, the inflammatory response, a defence used by fish against invading disease organisms, is suppressed by hormones released from the adrenal gland.

Water balance in the fish (osmoregulation) is disrupted due to changes in the metabolism of minerals. Under these circumstances, freshwater fish absorb excessive amounts of water from the environment (over-hydrate); saltwater fish lose water to the environment (dehydrate), This disruption increases energy requirements for osmoregulation. Respiration increases, blood pressure increases, and reserve red blood cells are released into the blood stream.

Fish are able to adapt to stress for a period of time; they may look and act normal. However, energy reserves are eventually depleted and hormone imbalance occurs, suppressing their immune system and increasing their susceptibility to infectious diseases.

Defence against Infection: Mucus

Mucus (slime layer) is the first physical barrier that inhibits entry of disease organisms from the environment into the fish. It is also a chemical barrier, containing enzymes and antibodies which can kill invading disease organisms. Mucus also lubricates the fish, aiding their movement through water, and is important for osmoregulation.

Injury as a result of handling (i.e., capture, transport, etc.) and certain chemicals in the water (i.e., poor water quality, disease treatments) remove or damage the mucous layer, reducing its effectiveness as a barrier against infection at a time when it is needed most. This damage decreases the chemical protection of the slime layer and also results in excessive uptake of water by freshwater fish and dehydration by saltwater fish. Decreased lubrication causes the fish to expend more energy to swim at a time when its energy reserves are already depleted.

Scales and Skin

Scales and skin function as a physical barrier which protects the fish. These are injured most commonly by handling, rough surfaces of tanks or cages, and by fighting caused by overcrowding or reproductive behaviour. Parasite infestations can also result in damage to gills, skin, fins, and loss of scales. Damage to scales and skin of the fish can increase the susceptibility to infection. It also causes excessive uptake of water by freshwater fish or loss of water from marine species (osmotic stress). Fish which are heavily parasitized may die from bacterial infections which gained initial entrance to the fish's body through damaged areas in the skin.

Inflammation

Inflammation is a natural immune response by the cells to a foreign protein, such as bacterium, virus, parasite, fungus, or toxin. Inflammation is characterised by swelling, redness, and loss of function.

It is a protective response, an attempt by the body to wall off and destroy the invader. Any stress causes hormonal changes which decrease the effectiveness of the inflammatory response.

Temperature stress, particularly cold temperatures, can completely halt the activity of the immune system, eliminating this defence against invading disease organisms. Excessively high temperatures are also extremely detrimental to the fish's ability to withstand infections. High water temperature may favour rapid population growth of some pathogens. High temperature also reduces the ability of the water to hold oxygen and increases the metabolic rate and resulting oxygen demand of the fish.

Antibodies

Unlike inflammation and other nonspecific forms of protection, antibodies are compounds formed by the body to fight specific foreign proteins or organisms. The first exposure results in the formation of antibodies by the fish which will help protect it from future infection by the same organism.

Exposure to sublethal concentrations of pathogens is important for fish to develop a competent immune system. Animals raised in a sterile environment will have little protection from disease. Young animals may not have as effective an immune response as older animals and therefore, may be more susceptible to pathogens in the environment.

Stress impairs the production and release of antibodies. Temperature stress, particularly rapid changes in temperature, severely limits the fish's ability to release antibodies, giving the invader time to reproduce and overwhelm the fish. Prolonged stress reduces the effectiveness of the immune system, increasing the opportunities for disease-causing organisms.

Disease Prevention

Numerous books and articles have been written on the diagnosis and treatment of specific fish diseases; however, prevention through good management practices is the best control measure to minimize disease problems and fish kills. Good management involves maintaining good water quality, preventing injury and stress during handling providing good nutrition, and using sanitation procedures, The following are management practices that help prevent stress and the resulting fish kills.

Water Quality

1. Do not exceed carrying capacity of fish in ponds and tanks.
2. Monitor water quality parameters.
3. Maintain dissolved oxygen levels above 5 mg/L. Sub-optimum levels of dissolved oxygen, while not immediately lethal, may stress fish, resulting in delayed mortality.
4. Prevent the accumulation of organic debris, nitrogenous wastes (ammonia and nitrite), carbon dioxide, and hydrogen sulfide.
5. Maintain appropriate pH, alkalinity, and temperature for the species.

Handling and Transporting

1. Use capture methods that minimize physical injury and stress.
2. When possible, use knitted mesh nets rather than knotted nets to reduce injury and scale loss.
3. Speed and gentleness when handling fish are of utmost importance.
4. Minimize the number of times the fish are lifted from the water, and work as quickly as possible when transferring fish.
5. Harvest, handle, and transport fish at times when fish are least susceptible to stress and infection.
6. Transport and holding tanks should be large enough to allow complete freedom of movement of fish and have no sharp corners or edges that might injure the fish.
7. Maintain optimum water conditions while capturing, hauling, and handling fish.
8. A high level of dissolved oxygen is crucial for rapid recovery of the fish from the struggle of capture and handling.
9. Salt (0.3 to 1.0 percent) maybe used in the transport water to minimize osmotic stress and bacterial infection of freshwater fish.
10. Ice may be added to the water during hauling to prevent an increase in water temperature which reduces the ability of the water to hold oxygen and increases the metabolic rate and resulting oxygen demand of the fish.

Nutrition

1. Feed a high quality diet that meets the nutritional requirements of the species.
2. Use proper feeding rate (either over-feeding or starving the fish should be avoided).
3. Store feed in a cool dry place to prevent deterioration. If available, a freezer is ideal for storing fish feed.

Sanitation

1. Quarantine all new fish and observe for mortality. Send samples to a diagnostic laboratory to be examined for parasites and evaluated for viral and bacterial disease.
2. Prevent disease-carrying fish from living in hatchery water supply (e.g., reservoir ponds, springs, streams).
3. Remove all dead fish from a production system as soon as they are observed.
4. Dispose of dead fish properly to prevent spread of disease.
5. Use good sanitation practices resulting in clean equipment, ponds, and tanks. Disinfect containers, nets, and equipment to minimize transmission of parasites and disease from one population to another.

Stress compromises the fish's natural defences against invading pathogens. When disease outbreaks occur, the underlying stress factors, as well as the disease organism, should be identified.

Correcting stress factors should precede or accompany chemical disease treatments. A disease treatment is only an artificial way of slowing down an infection so that the fish's immune system has time to respond. Any stress which adversely affects the fish will result in an ongoing disease problem. Prevention of disease outbreaks is more cost-effective than treating dying fish.

Stress in Fish

Stress in fish is an habitual, undesirable aspect of production. Stress results from biotic and/or abiotic challenges that act in changing or modifying the animal's natural or homeostatic state. Common fish stressors involve alterations in the fish's immediate environment such as:

- Chemical changes - contaminants, low oxygen and acidification
- Physical changes - handling, capture, confinement and transport
- Perceived changes - startling or predators

The above stressors are categorized as either acute or chronic depending on their duration and frequency. Acute fish stressors are events which the animal experiences for a short period of time such as handling. Chronic fish stressors are defined as a constant or recurring exposure that causes a prolonged physiological response. Fish stressors of both types have severe negative effects on the animal's growth and health.

The Physiological Response to Stress in Fish

Generally, stress in fish is a non-specific response classified as three distinct responses:

Primary Response

The primary response is characterised by the activation and the secretion of the hormones, corticosteroids (cortisol) and catecholamines, into the blood.

Secondary Response

The release of hormones triggers the secondary response which involves the release of glucose into the blood for energy production, followed by increases in heart rate, gill blood flow and metabolic rate which causes changes to blood lactate and hematocrit.

Tertiary Response

Changes in blood physiology ultimately cascade into a whole body change or the tertiary stress responses. Changes associated with the tertiary stress response include reduced growth rate, decreased disease resistance, altered behaviour and reduced survivability.

The Anaesthetic Process

Fish anaesthesia is a continual process that ranges from a slight loss of sensation through to medullary collapse (or death). It is generally accepted that there are five stages of anaesthesia in fish, however for aquaculture operations it is more relevant to refer to three states of anaesthesia: sedation, handleable and deep anaesthesia.

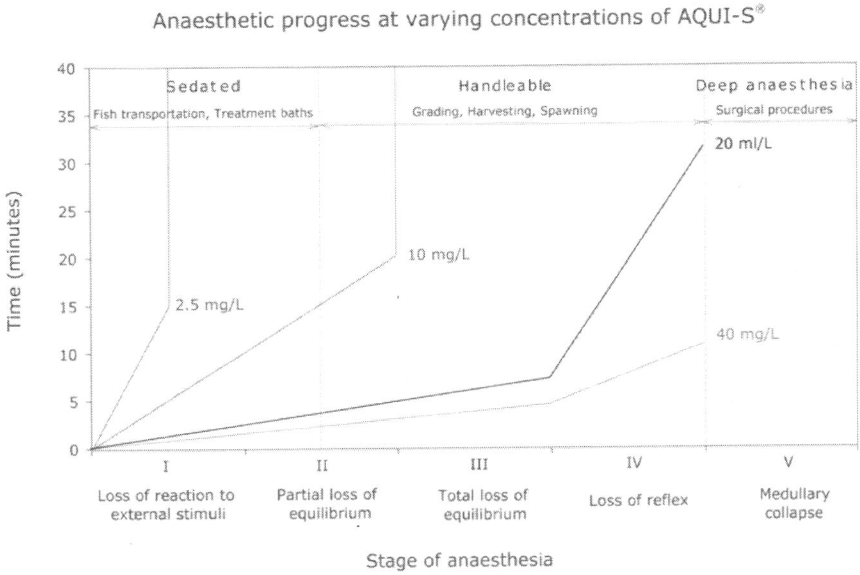

How Fish Proceed to Deep Anaesthesia?

Sedation: At low concentrations of anaesthetic, as with the 5 mg/L above, a sedated state is reached by the animal and a condition exists where the metabolic clearance rate of the anaesthetic equals the rate of uptake by the animal. In this case the animal will remain sedated and never become fully anaesthetised.

Steady State Sedation: Anaesthetic uptake = Metabolic clearance

This steady state is ideal for transport operations where the duration of the treatment may extend for several hours.

Handleable

At a range of concentrations where the rate of uptake of the anaesthetic is equal to or greater than the rate of metabolic clearance, as with 10 mg/L above, the animals are deemed handleable. At these concentrations the animal looses equilibrium and may slowly progress deeper into anaesthesia.

Progression to anaesthesia: Anaesthetic uptake e" Metabolic clearance

A concentration that induces handleable animals is ideal for harvesting, grading and stripping operations as the animal shows no reaction to being handled.

Deep Anaesthesia

Above a certain concentration of the anaesthetic, as in the 20 mg/L example above, there is a steady progression from loss of reaction to external stimuli, through to loss of equilibrium, then loss of reflex and eventually to total medullary collapse. At higher concentrations, e.g. 40 mg/L, the time to progress to deep anaesthesia is shorter.

Progression to anaesthesia: Anaesthetic uptake > Metabolic clearance.

As the concentration is increased the handleable window for the aquaculture operator is reduced. Most fisheries operations will be carried out within the handleable window.

Chapter 3

Fish Intelligence and Communication

Being underwater, the fish's basic "intelligence" and sensory skills are adapted very differently from the land-based animals. The animals' primary concerns are, fundamentally, to eat and not be eaten. Aquatic dwellers also employ their senses mainly for self-preservation and the propagation of the species.

Fish sense hunger, pain, and threat. They become aggressive when biological triggers signal the craving for food, or the need to defend a brood or a territory. Fish also communicate with one another, and may even try to communicate with you. In order to begin to understand what your fishes are "saying," here's an introduction on how they function and interact with the environment.

Sight

Optimized to see food, partners, and potential predators, different types of fish eyes have evolved to adapt to the environment. Nocturnal fish have large eyes; muddy river dwellers have small eyes; cave dwellers are totally blind.

The eyes of mid-water fishes are usually located on the sides of their heads to give them a 360-degree view of everything around them. The eyes of bottom dwellers, on the other hand, are usually located on the top of their heads to catch all the action above them; while those of surface feeders have evolved into split focal vision to see both above and below the waterline.

Shortsighted compared to terrestrial creatures, most fish have monocular vision. This means that each eye can focus on an object independently of the other eye, affording them an all-around panoramic field of vision. On the other hand, most predatory fishes have binocular vision with eyes close to one another and in front of the head which, like in humans, allows an accurate assessment of the exact location of their prey.

Some shallow-water fish are equipped with false eyelids that they can lower to shield their eyes from bright sunlight. The majority of fish, however, do not have the benefit of these "shades." Because of this, it normally takes them a while to adjust to sudden changes in lighting, and it is best to wait a few minutes after turning the aquarium light on before feeding them.

As for colour vision, research has confirmed that fish can see colour, to some extent, above and beyond some practical uses of colour. For example, the fact that most strikingly coloured species use their pigmentation to attract mates; or use colour to hide from and ward off predators.

Hearing

Sound travels more than four times faster through water than through air. Even without a middle and outer ear like other animals, a fish can "hear" sound through the vibrations in the water, via a porous lateral line along the sides of its body. By picking up water vibrations, the fish becomes aware of feeding frenzies or fearful fleeing that happens nearby.

Aside from this sensory lateral line, most fish have an inner ear, called an otiolith, which functions very much like the human eardrum. It is usually located at the back of the skull, and like the human inner ear, it also helps the fish maintain orientation and balance through cilia that send messages to the brain.

Therefore, because fish can sense the smallest vibrations through the water, tapping the aquarium tank, the stand that holds it, or otherwise causing sound waves to travel through the water, can cause a great deal of shock, stress, and torment to the resident fish.

Touch

The sense of touch is highly developed in fish. This is not because it helps them differentiate textures or express emotions like other animals, but because it is closely associated with their sense of taste, direction, self-preservation, and territoriality.

Gouramis and Angelfish use their pelvic fins to taste and grope for food, but they also use their feelers as warning devices when coming out of a hiding place. They use their fins to feel their way around and, at times, to assert dominance during a territorial dispute. Catfish and most nocturnal species use their barbels to sense their way in the dark, to dig through the substrate, as well as to taste their food.

Taste

A fish's taste buds are located in receptors that are spread on several areas of the body. Many fish have receptors on their heads, mouths, lips, and on special extensions, like the barbels on the catfish. Many species have taste receptors on the skin, which sends food messages to the brain, and prompts the fish to swim towards the source of the food taste.

Goldfish have taste sensors inside the mouth – when the brain interprets a particle as inedible, the object is quickly expelled.

Smell

Many varieties of fish have nostrils with which to smell odors emanating from food and mating partners. These nostrils have sensors that detect odors from the water and send signals to the brain. Some of the more sensitive fish, like catfish, and those with poor eyesight, have connecting pairs of nostrils that allow water to pass into the front nostril and flow out the rear opening.

Not only do these sensors help the fish taste and zero-in on the food, they also function as receptors for the "odors" that some fish emit into the water to denote position and rank within a community or territory. Most females also emit scents when they are ready to mate.

Be Sensitive to Fish Sense

Aside from behaving in response to the natural stimuli in their tank environment, fish have also been known to respond to their keepers. Larger species, like the Oscar, Pacu, Grouper, Arowana, and Triggerfish, are receptive to hand feeding; they express hunger by following the movements of their owners.

If you are observant, you will also notice that fish behaviour can alert you to impending problems in your tank. Peaceful fish that suddenly show aggressive behaviour towards tank mates and aquatic foliage may be signalling stress or insufficient feedings. Territoriality may signal that a fish has reached maturity and must, therefore, be given a new set-up. Unusual behaviour, like hiding, refusing to eat, scratching, and erratic swimming are strong indications of health problems.

Your awareness and concern for your fishes instincts, senses, and means of communication will ensure that the community remains healthy and viable. Here is a summary of reminders:

1. Provide the appropriate lighting for all your fish. Avoid sudden changes of light that could shock or stress, especially the species that are nocturnal or highly sensitive to illumination.
2. Never tap the aquarium glass. Sound waves are magnified by water and the noise can be stressful to the fish.
3. Provide the appropriate substrate for fishes that use barbels or extended fins to taste or grope through the ground. Jagged gravel and sharp rocks can damage the fish's feelers and taste receptors.
4. When transferring fish from one location to another, use a soft net to minimize harm or damage to the fish's skin, fins, or orifices. Any damage will affect the fish's natural capabilities to sense and respond to its environment.
5. Be attentive to "body language" messages that your fish may be communicating, and respond accordingly.

Chemical Communication in Fishes

his piece and two following on Pheromonal Behaviour Effects and Growth Inhibiting Substance(s) of Fishes are intended as companion articles to the magnum dopus, A Review of Literature on Hormonal Manipulation of Fishes as an Aquacultural Technique (FAMA 6/92). These are on-going attempts at bringing to our aquarist/hobbyist some more baseline knowing of physiology as it relates to practical keeping and breeding.

The auspicious Paul Speice(ous) provided the muse again in reminding this humble writer of the need for re-re-reminding the uninitiated of the value of frequent partial water change as a primary function of sound system management. If you can get your hands on the industry pub. Pet Product News, vol.5 no. 9, Sept. 92, "The Aging Aquarium" (Fancy Publications Inc., P.O. Box 6050, Mission Viejo, CA 92690-6050, 714-855-8822, Fax 714-855-3045, free to the trade), do so. So, ahem, on with it!

Let's review a bit: this periodical is aimed at enhancing/reinforcing our experience-enjoyment of ornamental aquatics, any and all fields/ ways of knowing aquatics directly contributes to this ends. Hence my argument for technical understanding; science/engineering.

The scientific (testable, falsifiable hypothetical) study of life is termed Biology. Within this body of tentative facts, ideas, methods, attitudes and techniques is a sub-field of study called Physiology; scientific study of metabolic function. Even within this area are specialities of interest. One is called Endocrinology ("inside" "cry""study"); the scientific study/practice involving the physiology of hormones (internal chemical messengers).

This and the next couple of articles discuss topics in a related category exocrinology, the science of chemicals that are "cried outside". These are simple to complex compounds, some of tremendous importance to us as aquariculturists, profoundly affecting growth, reproduction, colour, immune systems (generally all aspects of health). These externally released chemical messengers are variously classified as eliciting behaviour of the same species (pheromones, "Fair Oh Moans") and other than the releasor's species (allomones, "allo" like aloe vera).

In this offering we'll journey together on the issue of chemical communication in fishes, what it is, why you might/should be interested, and what you will want to do to optimize your situations. The following articles are more specific, concerning Growth Inhibiting Substance(s) affects, and Alarm Substance studies.

Chemical Communcation in General: It makes sense that substances released by aquatic organisms migh serve to regulate, instigate and moderate the behaviour of the same and other species. Not to be too reductionistic, but, as you know, 1) "water is the universal solvent" with more things being more soluble in it than any other material; 2) even with our senses of olfaction/gustation, nothing happens until whatever stuff dissolves into our fluids (mainly water) and gets into our systems' via aqueous solution. Many fishes have good vision and many do not. Same thing with hearing (remember this). Other senses have/are becoming evidently very important. The low frequency vibration sense of the lateralis systems and electromagnetic perception of cartilaginous (sharks, rays, skates, chimaeras) among other groups through the so-called pit organs, the Ampullae of Lorenzini are two of my favourite examples, that have become very highly attuned. External chemical sense is likewise been demonstrated to be extremely sensitive in some species.

Though most scientific studies have centred on work with sticklebacks, salmonids, american catfishes (ameiurids/"ictalurids") and sunfishes, there is growing awareness of how important chemical communication is among pet-fish. Especially the families grouped within the Superorder Ostariophysi, the catfishes, minnows and their relatives (barbs, danios, rasboras, koi and goldfish, loaches, hillstream fishes like chinese algae eaters), Characoids (tetras, hatchetfish, pencilfish,...), including many important captive hobbyist species, have a proven highly sophisticated chemical signalling system for clueing their school mates of present danger.

What are They: Everything from metabolites like ammonia to simple amino acids to oligo to polypeptides (few to several amino acids branched together), short-chain (8-12 carbon) fatty acids, and more have been detected to have discernible effects on the behaviour and physiology of fishes. These are produced within the host fish's body and released through excretion, secretion and rupture of the body wall via the gills, skin slime, urine, feces, or trauma.

Site of Reception: Is mainly olfactory. Most fish nostrils are used for smell, not breathing. Also, olfaction in some fishes is not just in the nostrils, but the mouth and body surface.

Experiments with eels have shown them to register response to as low as ten to the minus eighth (10-8) concentration of some chemicals. That's getting down to about one molecule in their nose period!

Importance to the Fish and To Us: Since the fifties, we have known that chemical communication functions in attraction and recognition of individuals, the opposite sex offspring or parents. Chemical communication maintains hierarchies, flight, schools, especially at night and guiding migrating fishes. Interesting experiments can/have been done with american bullhead catfishes and cichlids using diluted water from alpha/beta individuals, ablation and training/discrimination work showing that individuals and groups within a species do recognise each other and respond, at times overtly, to the chemical presence of conspecifics (members of the same species).

For hobbyists, it is of iminent importance to know and use this information for optimizing results. It turns out, you probably know the gist of what to do. Absolutely clean (e.g. sterile, or chemically inert) conditions are not practical or generally desired.

That is to say, that some of this "chemical soup" we've been referring to is beneficial, but not too much, or too selective. The universal rule is once again FREQUENT PARTIAL WATER CHANGES. Different rules of thumb apply depending on size of the system, filtration, feeding, et cetera rei, but do them! At least semi-religiously vent a percentage of the systems water and solids and replace with new.

I'd like to make mention of the importance of plant hormones (phytohormones), and yes, even invertebrates, in the overall health/ balance of a system. The closer you look, the more there seems to be.

Closing: The three most important set of factors affecting fish growth and reproduction are light, temperature and metabolite build-up. Perhaps this last group should be re-described as the presence/ absence of chemicals from all sources affecting the physiology of the target organisms, including chemical communication compounds.

Pheromones

Pheromones are chemicals released by living organisms that send information to other organisms of the same species via scent. These pheromones are released in response to stress, alarm, danger, and sexual fertility. They are released by both insects and mammals in many situations.

Insect Pheromones

Alarm Pheromone: In response to danger or stress, ants release a chemical signal that is sensed by other ants nearby. In response, these nearby ants travel towards the scent and work quickly in order to help their endangered nest mate. Other insects, including honeybees, have these alarm signals that are released via pheromones.

Trail Pheromone: In addition to releasing scents in response to danger, ants also release a scent when they are returning to their nest with food. This pheromone signals to other ants that there is food at the nest. As more ants follow this scent, it is renewed in order to tell more ants. As food quantities decrease, ants release a different scent that tells the others that there is no more food.

Mammal Pheromones

Releaser Pheromone: Releaser pheromones bring forth specific behaviour. Many mammals release pheromones that signal to others that they are in their "territory." This is seen in dogs urinating and also in many other mammals. There are also other reasons for releaser pheromones. For example, female rabbits and some other mammals release a scent that tells their young that they are ready to nurse. This scent then causes their young to immediately begin nursing.

Primer Pheromone

Primer pheromones cause a shift in the endocrine system of the animal receiving it. Many mammals, including rats and mice, release pheromones that cause sexual behaviour in the other sex. When at peak fertility, females release a scent that attracts a male and causes him to become aroused and exhibit mounting behaviour on the female, who is also exhibiting lordosis. The pheromones are sensed by the vomeronasal organ (VNO) in the brain. Another example of a primer pheromone is in rats. Female rats mature faster when exposed to the scent of an adult male rat. Also, the scent of a male rat of one strain can induce miscarriage in a female rat pregnant by a male of a different strain.

Information Pheromone: Information pheromones provide information about an animal's identity. By smelling another animal of the same species, they are able to tell what the animal ate last, if they are in heat, their level of dominance, and their level of health.

Schreckstoffes: Alarm Substances of Fishes

There are many anecdotal references in pet fish and industry literature as well as scientific experiments pointing up the presence of chemical messengers enhancing/stimulating fight/flight reactions of fishes. The sight of another individual can be proved to effect a fish's metabolism/behaviour, but even a very small amount of material from another member of the same species or just water from their system can elicit an extraordinary overt reaction on another individual. As keepers of aquatic life these elements of chemical communication are extremely important to our success, especially in the transport of same-species schools in small volumes of water.

What: Likened unto the "alarm calls" of birds, zebras and monkeys, alarm substances of fishes are ecto- or social hormones (pheromones) communicating the presence of danger to conspecifics (members of the same <or closely related> species).

Pandey retells a history of as Von Frisch (1938) describing his accidental discovery of a fright reaction in a European minnow Phoxinus phoxinus . He noted that on introduction of an injured fish the school dispersed, retreated and hid. Von Frisch further discerned that this fright substance (schreckstoff) was deriveable only from the fish's skin extract. Subsequently, other investigators have been busy studying the distribution of alarm substances and fright reactions in fishes.

These inquiries have mainly centred on the important pet fish SuperOrder Ostariophyi (minnows <barbs, danios, rasboras, goldfish, koi, "sharks">, tetras, catfishes, many knifefish families (the suborder Gymnotoidei)), and two of the important food-fish orders Gonorhynchiformes (milkfish), and Salmoniformes (salmons and trouts, of course). These pheromonal behaviour effects have now been demonstrated in virtually all living orders of fishes.

Where: Alarm substances have been isolated from excretions, body slime and most-celebratedly, in large, specialised goblet or club cells (Kolbenzellen) located in the skin of fishes in the principal groups listed above. These cells do not open to the surface, but rather release their contents after mechanical injury.

Why: An elaboration of the adaptive significance of alarm pheromone systems has been elucidated by Pfeiffer (1977). Some possible advantages are: 1) alerting members of the same species (conspecifics) to the presence of a predator, or away from a hunting spot, 2) reduction of cannibalism of young; that is, a strong fright reaction elicited by eating young of it's own species.

How: These chemicals are secreted or otherwise released into the environment where they alter the behaviour and/or physiology of other's by ingestion, absorption, taste (not just through the mouth), or smell.

Other Related Minutiae: There is the opposite of fright contagions released by the same species as well, id est chemicals that incite gathering, such as group feeding behaviour, and as they say on T.V., but wait, there's more...

Chemicals from other species may also elicit a fright reaction. The classic bear sticking a paw in upstream water and scaring away salmon is an okay example. My favourite, though I lifted from a Wall Street Journal article, concerns the Prez, George Bush and his lack of fishing success. Seems like George is of a type of human that secretes a large amount of the amino acid L-serine which is very repugnant to fish. The head of state would do well to bring along a professional bass fisherperson who secretes very little L-serine, as baiting his own hook assuredly sends fish scattering.

Salient Points: Alarm Substances probably represent the most important aspects of chemical communication in fishes. Their presence and effects in crowded closed systems with inadequate filtration almost certainly leads to tremendous negative stress and "anomalous" loss of growth, reproduction, vitality, even life. These compounds are found in and have been demonstrated to have significant behavioural and physiological consequences in the whole "evolutionary" range of fishes, lampreys, sharks to flatfishes (pleuronectids). The home hobbyist, industry professional, and large-scale aquaculturist would do well to engineer their facilities and/or arrange maintenance procedures to minimize negative effects of secretory, excretory and damage-source alarm substances.

The Functions of Body Slimes of Fishes

What Makes a Fish a Fish? Fins, Scales, Gills, Living in the Water?

Surprisingly enough there are fishes without true fins or scales whose respiration does not rely mainly on "gills", and those that spend a great deal of time out of the water. One trait that all living fishes do have in common however, is that they have body slimes, a mucoid covering on their very outsides. This trait is very important, particularly for its protective properties.

First off, where does this slime come from? The answer is that it originates from dispersed glandular cells, uni- or multicellular in the fish's epidermis. The type and displacement of these glands is of importance in classification. They produce a glycoprotein called mucin which, when mixed water, produces mucus. As an example in the extreme, consider a hagfish (closest family to the Lampreys), Psychedelic Gobies (aka Mandarins, family Callionymidae), or some species amongst the true eels. Hagfishes are particularly slimy; one of their common namses is "slime eel" though they are not closely related phylogenetically to the eels.

One method of capturing Hags is to fill a steel drum with fish heads and/or other offal, puncture this drum and lower it over a boat side on the continental shelf where these fishes are found. Hagfish will squeeze in to the holes in the punctured drum and eat so much they can't squeeze back out when the drum is hauled to the surface. Once captured, there is difficulty preserving the catch.

First they must be cleaned of the copious amounts of slime they produce. If you can grab one well enough to stick it in a bucket of clean water, this medium quickly becomes slimy as well. These animals possess large (pea-sized) multicellular slime glands. Ultimately, what has been done to prepare Hagfishes for preservation is to place a batch of these fishs in a hapless researcher's washing machine with jumbo amounts of enzymatic detergent on a continuous rinse cycle. This finally results in a slime-free hagfish ready for alcohol or formalin preservation.

General Structure of the Skin: The skin of fishes, like that of all vertebrates (amphibians, reptiles, birds, mammals) consists of two principal layers:

1) Superficial epidermis, and
2) Deeper dermis.

The epidermis in turn consists of two or more layers. The deepest is a series of close-paced, discrete cells called the germinal layer, or stratum germinativum. The outer cells are formed of its daughter cells. There is much variation in the outer cells, depending on the group of fishes being investigated. Body slimes are the products of these daughter cells and their degradation and, as such, are continuously replaced.

The dermis consists of thick connective tissue made up of two basic layers. It is thicker and more stable than the epidermis.

What the Slime Does for Fishes: The fact that all fishes have these body coverings is some indication of their importance. With either too much or too little of this slimy coat, any fish will soon die. The slime serves three functions for all fishes. It aids in:

1) Osmoregulation/Gas Transport: Slime provides a selective interface to maintain internal/externail ionic balance. One of the reasons freshwater fishes are constantly urinating is their bodies are "saltier" than the water around them and they tend to absorb water. The fish gets rid of this excess water by elimination. The opposite rationale applies to saltwater fishes. In addition to salt balance, the slime plays important roles in dermal respiration. Fish breathe through their skins, as do humans. If the amount or quality of the slime changes, it effects the efficiency of gas transport through the skin.
2) External Protection: Body slime prevents attachment of ectoparasites by making the surface of the fish slippery, sloughing off with the parasite and suffocating pathogens. It also acts as a bandage by covering over a wound caused by trauma or infection. Usually fishes with poorly developed scales are more slimy, for example, Characins (some known as Tetras) and their relatives.
3) Reduces Turbulence: Especially in fast-moving fishes the drag resulting from small spaces between scales and projecting body parts accounts for considerable energy loss (up to 30% by some estimates) in locomotion. The slime acts to smooth out these gaps.

Additional Functions of Body Slime: In addition to the abouve functions, many groups of fishes benefit in other ways from their body slimes. For some they help:

1) Coagulate Particles: Providing clean water in the immediate area around the fish, thus improving movement and dermal respiration. Some filter-feeding fishes pass the mucus forward into their mouths and eat it. For example, some of the Wrasses.
2) Produce Toxins: For example, some of the previously mentioned Hagfishes (family Myxinidae), closely related to the Lampreys (Petromyzontidae) that have ruined fisheries in the Great Lakes, immobilise a host on contact with their body slimes, entering their vent and eating them. *Pardochirus marmoratus*, a Sole (type of Flatfish) in the Red Sea featured in the November 1974 issue of National Geographic Magazine, has a slime that

contains a substance so effective in warding off shark bate that the attacker's jaws are said to be frozen in mid-bite.

3) Cocoon Formation: The African Lungfish avoids desiccation during Summer and dry periods by making a shell of its body slime and "hibernating".

Many Parrotfishes (family Scaridae) produce a mucus "tent" at night to protect themselves against predation. As an experiment, some of the Parrotfishes of the genus *Scarus* which construct such sleeping bags and an equal number of similar species of the similar-appearing genus *Sparisoma* which do not produce cocoons were placed in a tank with a few large Moray Eels (family Muraenidae). It seems Parrotfishes are a favourite food item of many Morays. These fishes were left together overnight. The *Scarus* built cocoons and were not eaten but the *Sparisoma* were consumed. During the night the Morays were observed approaching the camouflaged *Scarus*: although they couldn't see what was contained within the enclosed envelopes, it was evident they understood what was contained within. The eels "tasted" the mucus and left the *Scarus* alone.

Feeding: Several fishes, including some of the *Mystus* (Asian catfishes) and the Discus (*Symphysodon*) secrete body slimes to feed their young. Baby Discus feed on an overabundance of slime which develops on the sides of the parent fish at breeding time. The substance is highly proteinaceous in nature and is produced by specialised skin cells. This situation is not the same as lactation in mammals; the slime is different chemically and there is not permanent organised structure for secretion. This is an important source of food for the young who need it during the first week of life. There are no suitable naturally occurring substitutes.

Alarm Substances: A lot of aquarium fishes such as Tetras, Barbs, freshwater "Sharks", Rasboras, Loaches, Catfishes and others have a number of blind cells; that is, they have no opening to the outside of the body, that are involved in producing, storing alarm substances. When the skin is broken, these cells release a fright contagion that notifies others that something is going wrong. These substances are not necessarily species specific. They are responsible for producing the fright syndrome German aquarists refer to as "shrekstoff". A situation that hobbyists everywhere should be aware of and guard against, through careful netting, handling practices, and adequate filtration and maintenance.

Nest Building Materials: In some species such as the Gouramis (family Anabantidae) and Bettas, slime is utilised in the construction of "Bubble Nests" that males spit eggs and keep young safe in till they're able to fare on their own.

Cement: In western India the mucus of Snakeheads (family Channidae) is used in the construction industry to increase the strength of mortar.

Practical Aquarium Significance: What does all this mean to an aquarist? As can be seen from the previous discussion, body slimes are eminently important to fishes. Stress to the fish can and does occur by affecting body mucus amount or viscosity, and vice versa.

Aquarists should be especially careful when netting their fishes. Commercially, we never touch fine scaled fishes with our hands. If a fish drops to the floor, pick it up with a wet net or towel and try to preserve the integrity of the animals slime layer.

Metallic ion medications among other types act as proteinaceous precipitants, making the fish produce more slime as irritation increases. Concerning the fishes, the copper ions (as well as malachite green) sold as marine and freshwater ich remedies act as an irritant to the skin and gill membranes of fishes, which in response produce copious amounts of mucus to protect these tissues.

If disease organisms are present on the gills and skin, the mucus produced engulfs the organisms. When the mucus is sloughed off, the disease organisms are lost as well. Under high dosages or prolonged treatment with such medications, a loss of fish livestock results from direct uptake of medicants and mucoid production so great as to impede gaseous exchange by the gills and skin.

Much of these free metal ions can be found even in fresh tapwater. Some water treatment products are designed to make aquarium fishes more slimy to protect against such irritation.

Once again, the best policy for maintaining aquatic life is not to change too much too quickly and provide the optimum suitable environment. Change part of your water frequently, vary the diet and watch what you put into your tanks. A factor as seemingly "simple" as fish sliminess can be a determining factor in the well being of our aquatic charges.

Growth Inhibiting Substance(s) of Fishes

Do fish grow only as large as their genetic capability and physical environment allows? Well, sort of. Metabolite effects, are often more rate/size limiting. Say Whaaa?

Under my files of least favourite "here we go again" pet-fish urban myths to re-hash and debunk is, "are there growth inhibiting/ stimulating substances?" Yes, yes, and yes, and yes! they are important and interesting as all heck. Let's tackle what G.I.S. and G.S.S(ubstances) are, what they do to and for your wet pets, what you can/should do about this new or reinforced knowledge. Not enuf angst? Maybe next we'll tackle whether to float new introductions or not. (Hah!) Read on Bartholomew.

En media res (in the middle of things, eh?) Introduction: Geez, it just seems like (yesterday) the latter 60's and 70's the topic of whether fishes, and other living critters, plants, algae... produced chemicals that influence positively and/or negatively their growth (reproduction, colour,...) was being heavily disputed, even disparaged. Allow me to call-up the names of some of my hobbyist bulletin writer-heroes, Jim (the hammer) Langhammer and Kappy Sprenger, among others from those times to present for clear & fun cerebral activity on this and many other issues.

All living things, by definition, engage in biological, read that as bio-chemical/physical reactions transforming some materials, utilising energy and producing "waste" and growth products and so-called by-products. The various chemicals changed and passed outside particular species may be classified in the following way:

1) Having no specific/known effect(s) on the secreting species or others.
2) Having such effects:

A) On the species itself (pheromones)

B) On other species (allomones)

What is this Stuff, Anyway?

Many exocrine (versus endocrine) substances have been isolated/ identified to have G.I.S. and/or G.S.S. effects. Ammonia-derived metabolites, amino acids and combinations of A.A.s, and most notably short chain fatty acids, have been demonstrated to accelerate/retard the individual growth and behaviours of many groups of fishes.

Most "advanced" (old and wrinkled?) aquariologists know the practical whys and wherefores (consequences) of these substances. Some highlights:

1) Everything else being equal (that is, no food, gas, physical space, filtration, lighting, temperament, other limitations) some same-species fish substances:

A) restrict spawn sizes, differential growth rates of individuals and groups of individuals, further:

B) limiting their subsequent reproductive viability, and even:

C) the smaller (runt) individuals survival rate (demise).

2) Mixing species often lessens the pheromonal stunting effects of a single species alone.

3) Boiling the water, leaving it to age (for weeks), dissipates the responsible compounds.

4) Removal of some or all of the larger members of a spawn or population spurs the growth/development of the next larger individual(s).

5) Some of these materials have beneficial and antagonistic impacts. That is, using "old water" has shown healing and growth-positive actuation compared with absolutely fresh (i.e. no-same-species-chemical secretions) water

6) That frequent massive to partial water changes and effective chemical filtration lessen these effects.

Why? Tell Me Why?:

Towards what possible ends would/should organisms produce such self-limiting secretions/excretions. Several reasons come to mind: 1) To preclude over-crowding and loss of the whole population due to food, oxygen/carbon dioxide limitations, other controlling factors, in the otherwise absence of other species competition or predatory pressures. And I'll leave it up to you to decide whether this one should have been mentioned first, 2) the ole Chucky Darwin Natural Selection (long may it wave?). By having the most humongous citizens beat out the less-humongous as a gauge of "fitness", may enhance the overall survivability of the species.

Where's This Stuff Come From?:

Principally growth stimulating/retarding substances have been found, and presumably are released by way of, body slime, excretory discharges from the anus and gills, and traumatic damage to the organism in general. This last category will explore in our next visit as Schreckstoffes: Alarm Substances of Fishes.

Closing: (Sort of):

So... What's an average pet-fish-ichthyologist to do, anyhow? Mainly, not worry too much; at least not more than their benefiting from their aquatic experience. Actually there are (to my understanding) only a couple or three "things to do" to ameliorate the cumulative negative effects of metabolite build up:

1) Try to selectively filter/change these compounds. Various carbons and clinoptilolites (e.g. zeolite) have showed various positive results in removal of ammonia salts. Still gotta plug skimmer/protein skimmers, with or without ozone et alia embellishments.
2) Serial dilution through, Yes, my favourite: FREQUENT PARTIAL WATER CHANGES; Yay! Certainly the best, least expensive, effective means. Oh, semi-lastly,
3) Flood the system with complementary chemicals. Live plants, a mix of communities of macro and micro-organisms, the whole "life'juice" of the system... the more complex and complete, larger, the better. Live plants, algae, generally all organisms investigated also have their phytohormones et al. affecting/influencing their own and other species. The more these are integrated and functioning, the more naturally homogeneous and self-stabilising (homeostatic) your system will be. Therefore the argument for linking your tanks together, use live plants, live rock, etc, and finally, lastly
4) The usual harangue about mis/over-feeding. The more glop tossed in, the worse. Feed sparingly, at correct intervals, of useful foods. Are you feeding for growth or maintenance or what?

The Smell of Fear: Ruffe "Alarm" Pheromones

Like most freshwater fish, Eurasian ruffe have a keen sense of smell. Sorensen, a pioneer of fish pheromone research, and his colleagues in the Department of Fisheries and Wildlife understand that smells can often be powerfully attractive or repulsive. They are attempting to harness the capacity of pheromones to help manage fish populations.

"Pheromones are chemical signals that pass between members of the same species that evoke potent, instinctual responses," said Sorensen. "Pheromones are environmentally-benign, highly-specific, and conceivably an inexpensive way to manage populations of nuisance fish."

During an investigation of attractant (sex) pheromones, Sorensen and his colleagues switched tactics and discovered that Eurasian ruffe repel other ruffe with an offensive odor, an alarm pheromone, that seeps from their injured skin. "Damaged ruffe skin emits an extremely repugnant odor that suppresses the swimming and feeding activities of exposed ruffe," said Sorensen. "Ruffe respond almost immediately and very dramatically to this alarm pheromone and, when possible, avoid it. Clearly, this cue has potential for managing ruffe. The key remaining question is how effective it will be in the large open spaces of the lake," he said.

Pheromones aren't just "smells." They are probably the most ancient form of animal communication. "Fish commonly use pheromone cues to navigate and coordinate intraspecific activities in waters that are often turbid, vast, and relatively featureless," said Sorensen. Pheromones are already common for controlling insects. Generally, insect pheromones are species-specific and the same seems true of fish pheromones. In Sorensen's study, ruffe did not respond to the alarm pheromones of injured yellow perch (a member of the same subfamily as ruffe) nor did goldfish respond to injured ruffe.

Although organisms can lose their sensitivity to a stimulus after long-term exposure, Sorensen doesn't believe wild Eurasian ruffe would normally become desensitized to their alarm pheromone; the Great Lakes are so huge and the initial response of the fish is to swim away. "There is no obvious reason why they would continue to expose themselves to it."

The perch-like Eurasian ruffe became part of the fauna in the Duluth-Superior Harbor in the early 1980's. Presumably, they crossed the Atlantic Ocean as accidental passengers in the ballast tanks of cargo ships. Within a decade of their detection, Eurasian ruffe became the most abundant fish trawled from the bottom of the harbor.

Currently, ruffe have spread along Lake Superior's south shore to the Fire Steel River, east of Ontonagan, MI; to several harbors along the north shore to Thunder Bay, Ontario; and to Alpena, MI, on Lake Huron. Attempts to control the expanding population of ruffe with predators and piscicides (chemicals that kill fish) have failed and alternatives are being sought.

Managing invasive fish species with their own non-toxic pheromones is a new and attractive idea. Authorities in Alpena have approached Sorensen about using the alarm pheromone to exclude ruffe from areas where they are not wanted, such as docks where ships take on ballast water. Sorensen and his coworkers also characterised a ruffe sexual attractant that might be used in conjunction with the alarm pheromone.

Glucocorticoid

Glucocorticoids (GC) are a class of steroid hormones that bind to the glucocorticoid receptor (GR), which is present in almost every vertebrate animal cell. The name glucocorticoid (glucose + cortex + steroid) derives from their role in the regulation of the metabolism of glucose, their synthesis in the adrenal cortex, and their steroidal structure.

GCs are part of the feedback mechanism in the immune system that turns immune activity (inflammation) down. They are therefore used in medicine to treat diseases caused by an overactive immune system, such as allergies, asthma, autoimmune diseases and sepsis. GCs have many diverse (pleiotropic) effects, including potentially harmful side effects, and as a result are rarely sold over the counter. They also interfere with some of the abnormal mechanisms in cancer cells, so they are used in high doses to treat cancer. This includes mainly inhibitory effects on lymphocyte proliferation (treatment of lymphomas and leukaemias) and mitigation of side effects of anticancer drugs.

GCs cause their effects by binding to the glucocorticoid receptor (GR). The activated GR complex, in turn, up-regulates the expression of anti-inflammatory proteins in the nucleus (a process known as transactivation) and represses the expression of proinflammatory proteins in the cytosol by preventing the translocation of other transcription factors from the cytosol into the nucleus (transrepression).

Glucocorticoids are distinguished from mineralocorticoids and sex steroids by their specific receptors, target cells, and effects. In technical terms, “corticosteroid” refers to both glucocorticoids and mineralocorticoids (as both are mimics of hormones produced by the adrenal cortex), but is often used as a synonym for “glucocorticoid”.

Cortisol (or hydrocortisone) is the most important human glucocorticoid. It is essential for life, and it regulates or supports a variety of important cardiovascular, metabolic, immunologic, and homeostatic functions. Various synthetic glucocorticoids are available; these are used either as replacement therapy in glucocorticoid deficiency or to suppress the immune system.

Effects

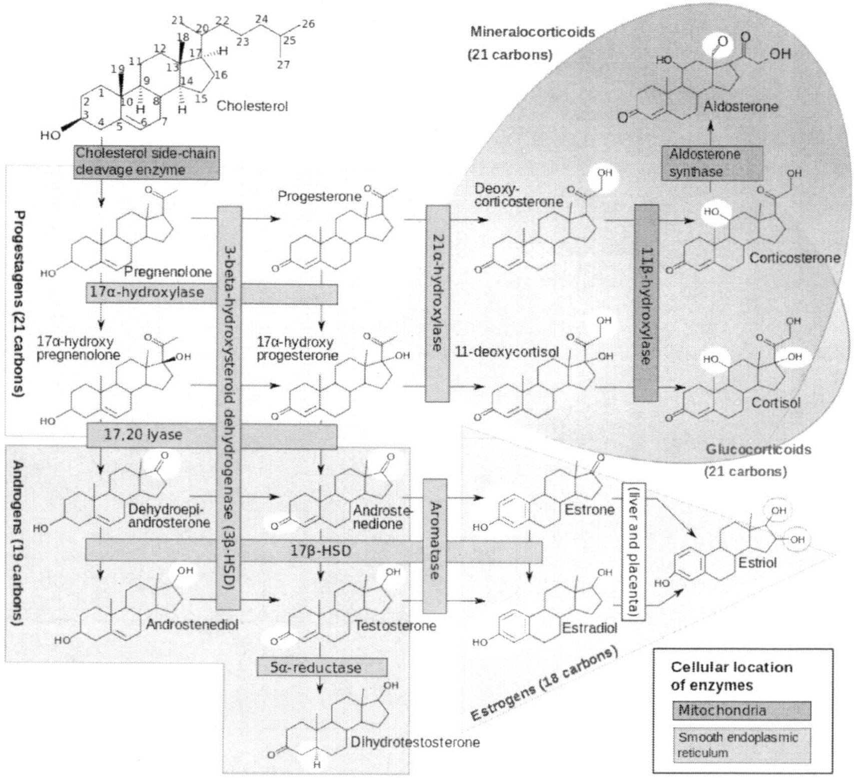

Figure: Steroidogenesis showing glucocorticoids in green ellipse at right - it is not a strictly bounded group, but a continuum of structures with increasing glucocorticoid effect, with the primary example being cortisol.

Glucocorticoid effects may be broadly classified into two major categories: immunological and metabolic. In addition, glucocorticoids play important roles in fetal development.

Immune

As discussed in more detail below, glucocorticoids function through interaction with the glucocorticoid receptor:

- up-regulate the expression of anti-inflammatory proteins
- down-regulate the expression of proinflammatory proteins

Glucocorticoids are also shown to play a role in the development and homeostasis of T lymphocytes. This has been shown in transgenic mice with either increased or decreased sensitivity of T cell lineage to glucocorticoids.

Metabolic

The name "glucocorticoid" derives from early observations that these hormones were involved in glucose metabolism. In the fasted state, cortisol stimulates several processes that collectively serve to increase and maintain normal concentrations of glucose in blood.

Metabolic effects:

- *Stimulation of gluconeogenesis, in particular, in the liver:* This pathway results in the synthesis of glucose from nonhexose substrates, such as amino acids and glycerol from triglyceride breakdown, and is particularly important in carnivores and certain herbivores. Enhancing the expression of enzymes involved in gluconeogenesis is probably the best-known metabolic function of glucocorticoids.
- *Mobilisation of amino acids from extrahepatic tissues:* These serve as substrates for gluconeogenesis.
- *Inhibition of glucose uptake in muscle and adipose tissue:* A mechanism to conserve glucose
- *Stimulation of fat breakdown in adipose tissue:* The fatty acids released by lipolysis are used for production of energy in tissues like muscle, and the released glycerol provide another substrate for gluconeogenesis.

Excessive glucocorticoid levels resulting from administration as a drug or hyperadrenocorticism have effects on many systems. Some examples include inhibition of bone formation, suppression of calcium absorption (both of which can lead to osteoporosis), delayed wound healing, muscle weakness, and increased risk of infection. These observations suggest a multitude of less-dramatic physiologic roles for glucocorticoids.

Developmental

Glucocorticoids have multiple effects on fetal development. An important example is their role in promoting maturation of the lung and production of the surfactant necessary for extrauterine lung function. Mice with homozygous disruptions in the corticotropin-releasing hormone gene die at birth due to pulmonary immaturity. In addition, glucocorticoids are necessary for normal brain development, by initiating terminal maturation, remodelling axons and dendrites, and affecting cell survival.

Arousal and Cognition

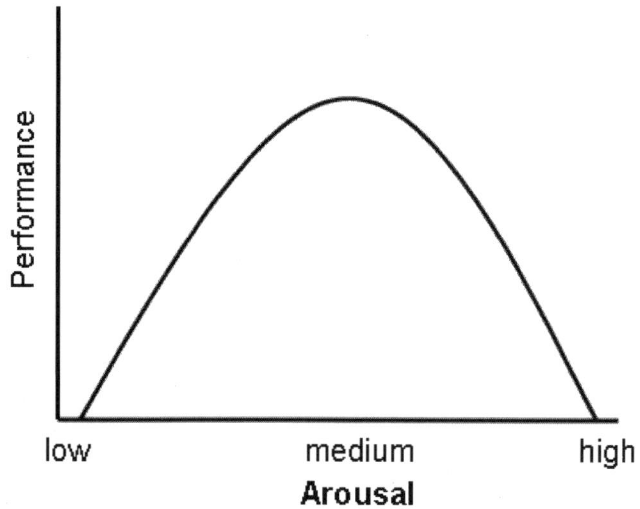

Figure: A graphical representation of the Yerkes-Dodson curve

Glucocorticoids act on the hippocampus, amygdala, and frontal lobes. Along with adrenaline, these enhance the formation of flashbulb memories of events associated with strong emotions, both positive and negative. This has been confirmed in studies, whereby blockade of either glucocorticoids or noradrenaline activity impaired the recall of emotionally relevant information. Additional sources have shown subjects whose fear learning was accompanied by high cortisol levels had better consolidation of this memory (this effect was more important in men). Glucocorticoids have also been shown to have a significant impact on vigilance and cognitive performance. This appears to follow the Yerkes-Dodson curve, as studies have shown circulating levels of glucocorticoids vs. memory performance follow an upside-down U pattern, much like the Yerkes-Dodson curve. For example, long-term potentiation (the process of forming long-term memories) is optimal when glucocorticoid levels are mildly elevated, whereas significant decreases of LTP are observed after adrenalectomy (low-GC state) or after exogenous glucocorticoid administration (high-GC state). Elevated levels of glucocorticoids enhance memory for emotionally arousing events, but lead more often than not to poor memory for material unrelated to the source of stress/emotional arousal. In contrast to the dose-dependent enhancing effects of glucocorticoids on memory consolidation, these stress hormones have been shown to inhibit the retrieval of already stored information.

Mechanism of Action

Transactivation: Glucocorticoids bind to the cytosolic glucocorticoid receptor (GR). This type of receptor is activated by ligand binding. After a hormone binds to the corresponding receptor, the newly formed receptor-ligand complex translocates itself into the cell nucleus, where it binds to glucocorticoid response elements (GRE) in the promoter region of the target genes resulting in the regulation of gene expression. This process is commonly referred to as transactivation.

The proteins encoded by these up-regulated genes have a wide range of effects, including, for example:

- anti-inflammatory – lipocortin I, p11/calpactin binding protein and secretory leukoprotease inhibitor 1 (SLPI)
- increased gluconeogenesis – glucose-6-phosphatase and tyrosine aminotransferase

Transrepression

The opposite mechanism is called transrepression. The activated hormone receptor interacts with specific transcription factors and prevents the transcription of targeted genes. Glucocorticoids are able to prevent the transcription of proinflammatory genes, including interleukins IL-1B, IL-4, IL-5, and IL-8, chemokines, cytokines, GM-CSF, and TNFA genes.

The ordinary glucocorticoids do not distinguish among transactivation and transrepression and influence both the "wanted" immune and the "unwanted" genes regulating the metabolic and cardiovascular functions. Intensive research is aimed at discovering selectively acting glucocorticoids that will be able to repress only the immune system.

Genetically modified mice that express a modified GR incapable of DNA binding are still responsive to the anti-inflammatory effects of glucocorticoids, while the stimulation of gluconeogenesis by glucocorticoids is blocked. This result strongly suggests most of the desirable anti-inflammatory effects are due to transrepression, while the undesirable metabolic effects arise from transactivation, a hypothesis also underlying the development of selective glucocorticoid receptor agonists.

Nongenomic

Glucocorticoids have been shown to exert a number of rapid actions that are independent of the regulations of gene transcription. Binding of corticosteroids to the glucocorticoid receptor (GR) stimulates phosphatidylinositol 3-kinase and protein kinase AKT, leading to endothelial nitric oxide synthase (eNOS) activation and nitric oxide-dependent vasorelaxation. Membrane associated GR has been shown to mediate lymphocytolysis. In addition, some glucocorticoids have been shown to rapidly inhibit the release of the inflammatory prostaglandin PGE2, this effect is blocked by the glucocorticoid receptor antagonist mifepristone (RU-486) and is not affected by protein synthesis inhibitors. These data together suggest a nongenomic mechanism of action.

Glucocorticoid-Induced Neutrophilia

Acute or chronic administration of corticosteroids causes neutrophilia, suggesting the enhanced release of PMNs from the bone marrow is an important mechanism of the glucocorticoid-induced granulocytosis. An alternative mechanism for the granulocytosis induced by glucocorticoids is an influx of PMNs from the intravascular marginated PMN pools. The response is caused by a shift of cells from the marginal to the circulating pool; hence, it frequently is referred to as demargination.

Some of the immunosuppressive effects of glucocorticoids are mediated by nongenomic signalling involving the glucocortiocid receptor (GR). A multiprotein complex composed of the unliganded glucocorticoid receptor, Hsp90, and the tyrosine kinases LCK and FYN is recruited to the antigen-activated T cell receptor (TCR) in T cells. This GR complex is necessary for TCR signalling. On binding of glucocorticoids to GR, this complex dissociates, thus blocking TCR signalling.

Pharmacology

A variety of synthetic glucocorticoids, some far more potent than cortisol, have been created for therapeutic use. They differ in both pharmacokinetics (absorption factor, half-life, volume of distribution, clearance) and pharmacodynamics (for example the capacity of mineralocorticoid activity: retention of sodium (Na+) and water; renal physiology).

Because they permeate the intestines easily, they are administered primarily *per os* (by mouth), but also by other methods, such as topically on skin. More than 90% of them bind different plasma proteins, though with a different binding specificity. Endogenous glucocorticoids and some synthetic corticoids have high affinity to the protein transcortin (also called corticosteroid-binding globulin), whereas all of them bind albumin. In the liver, they quickly metabolize by conjugation with a sulfate or glucuronic acid, and are secreted in the urine.

Comparative steroid potencies			
Name	*Glucocorticoid potency*	*Mineralocorticoid potency*	*Duration of action ($t_{1/2}$ in hours)*
Hydrocortisone (cortisol)	1	1	8
Cortisone acetate	0.8	0.8	oral 8, intramuscular 18+
Prednisone	3.5-5	0.8	16-36
Prednisolone	4	0.8	16-36
Methylprednisolone	5-7.5	0.5	18-40
Dexamethasone	25-80	0	36-54
Betamethasone	25-30	0	36-54
Triamcinolone	5	0	12-36
Beclometasone	8 puffs 4 times a day equals 14 mg oral prednisone once a day	-	-
Fludrocortisone acetate	15	200	24
Deoxycorticosterone acetate (DOCA)	0	20	-
Aldosterone	0.3	200-1000	-

Glucocorticoid potency, duration of effect, and overlapping mineralocorticoid potency varies. Cortisol (hydrocortisone) is the standard of comparison for glucocorticoid potency. Hydrocortisone is the name used for pharmaceutical preparations of cortisol. Data refer to oral dosing, except when mentioned. Oral potency may be less than parenteral potency because significant amounts (up to 50% in some cases) may not be absorbed from the intestine. Fludrocortisone, DOCA

(Deoxycorticosterone acetate), and aldosterone are, by definition, not considered glucocorticoids, although they may have minor glucocorticoid potency, and are included in this table to provide perspective on mineralocorticoid potency.

Therapeutic Use

Glucocorticoids may be used in low doses in adrenal insufficiency. In much higher doses, oral or inhaled glucocorticoids are used to suppress various allergic, inflammatory, and autoimmune disorders. Inhaled glucocorticoids are the second-line treatment for asthma. They are also administered as post-transplantory immunosuppressants to prevent the acute transplant rejection and the graft-versus-host disease. Nevertheless, they do not prevent an infection and also inhibit later reparative processes.

Physiological Replacement

Any glucocorticoid can be given in a dose that provides approximately the same glucocorticoid effects as normal cortisol production; this is referred to as physiologic, replacement, or maintenance dosing. This is approximately 6–12 mg/m^2/day (m^2 refers to body surface area (BSA), and is a measure of body size; an average man's BSA is 1.7 m^2).

Therapeutic Immunosuppression

Glucocorticoids cause immunosuppression, and the therapeutic component of this effect is mainly the decreases in the function and numbers of lymphocytes, including both B cells and T cells.

The major mechanism for this immunosuppression through inhibition of nuclear factor kappa-light-chain-enhancer of activated B cells(NF-êB). NF-êB is a critical transcription factor involved in the synthesis of many mediators (i.e., cytokines) and proteins (i.e., adhesion proteins) that promote the immune response. Inhibition of this transcription factor, therefore, blunts the capacity of the immune system to mount a response.

Glucocorticoids suppress cell-mediated immunity by inhibiting genes that code for the cytokines IL-1, IL-2, IL-3, IL-4, IL-5, IL-6, IL-8 and IFN-ã, the most important of which is IL-2. Smaller cytokine production reduces the T cell proliferation.

Glucocorticoids, however, not only reduce T cell proliferation, but also lead to another well known effect - glucocorticoid-induced apoptosis.

The effect is more prominent in immature T cells still inside in the thymus, but peripheral T cells are also affected. The exact mechanism underlying this glucocorticoid sensitivity still remains to be elucidated.

Glucocorticoids also suppress the humoral immunity, thereby causing a humoral immune deficiency. Glucocorticoids cause B cells to express smaller amounts of IL-2 and of IL-2 receptors. This diminishes both B cell clone expansion and antibody synthesis. The diminished amounts of IL-2 also cause fewer T lymphocyte cells to be activated. Since glucocorticoid is a steroid, it regulates transcription factors; another factor it down-regulates is the expression of Fc receptors on macrophages, so there is a decreased phagocytosis of opsonised cells.

Anti-Inflammatory

Glucocorticoids are potent anti-inflammatories, regardless of the inflammation's cause; their primary anti-inflammatory mechanism is lipocortin-1 (annexin-1) synthesis. Lipocortin-1 both suppresses phospholipase A2, thereby blocking eicosanoid production, and inhibits various leukocyte inflammatory events (epithelial adhesion, emigration, chemotaxis, phagocytosis, respiratory burst, etc.). In other words, glucocorticoids not only suppress immune response, but also inhibit the two main products of inflammation, prostaglandins and leukotrienes. They inhibit prostaglandin synthesis at the level of phospholipase A2 as well as at the level of cyclooxygenase/PGE isomerase (COX-1 and COX-2), the latter effect being much like that of NSAIDs, potentiating the anti-inflammatory effect.

In Addition, Glucocorticoids also Suppress Cyclooxygenase Expression

Glucocorticoids marketed as anti-inflammatories are often topical formulations, such as nasal sprays for rhinitis or inhalers for asthma. These preparations have the advantage of only affecting the targeted area, thereby reducing side effects or potential interactions. In this case, the main compounds used are beclometasone, budesonide, fluticasone, mometasone and ciclesonide. In rhinitis, sprays are used. For asthma, glucocorticoids are administered as inhalants with a metred-dose or dry powder inhaler.

Hyperaldosteronism

Glucocorticoids can be used in the management of familial hyperaldosteronism type 1. They are not effective, however, for use in the type 2 condition.

Resistance

Resistance to the therapeutic uses of glucocorticoids can present difficulty; for instance, 25% of cases of severe asthma may be unresponsive to steroids. This may be the result of genetic predisposition, ongoing exposure to the cause of the inflammation (such as allergens), immunological phenomena that bypass glucocorticoids, and pharmacokinetic disturbances (incomplete absorption or accelerated excretion or metabolism).

Side Effects

Glucocorticoid drugs currently being used act nonselectively, so in the long run they may impair many healthy anabolic processes. To prevent this, much research has been focused recently on the elaboration of selectively acting glucocorticoid drugs. These side effects could be prevented:

- immunodeficiency
- hyperglycemia due to increased gluconeogenesis, insulin resistance, and impaired glucose tolerance ("steroid diabetes"); caution in those with diabetes mellitus
- increased skin fragility, easy bruising
- negative calcium balance due to reduced intestinal calcium absorption
- steroid-induced osteoporosis: reduced bone density (osteoporosis, osteonecrosis, higher fracture risk, slower fracture repair)
- weight gain due to increased visceral and truncal fat deposition (central obesity) and appetite stimulation
- adrenal insufficiency (if used for long time and stopped suddenly without a taper)
- muscle breakdown (proteolysis), weakness, reduced muscle mass and repair
- expansion of malar fat pads and dilation of small blood vessels in skin
- anovulation, irregularity of menstrual periods
- growth failure, delayed puberty
- increased plasma amino acids, increased urea formation, negative nitrogen balance

- excitatory effect on central nervous system (euphoria, psychosis)
- glaucoma due to increased cranial pressure
- cataracts.

In high doses, hydrocortisone (cortisol) and those glucocorticoids with appreciable mineralocorticoid potency can exert a mineralocorticoid effect, as well, although in physiologic doses this is prevented by rapid degradation of cortisol by 11â-hydroxysteroid dehydrogenase isoenzyme 2 (11â-HSD2) in mineralocorticoid target tissues. Mineralocorticoid effects can include salt and water retention, extracellular fluid volume expansion, hypertension, potassium depletion, and metabolic alkalosis.

The combination of clinical problems produced by prolonged, excess glucocorticoids, whether synthetic or endogenous, is termed Cushing's syndrome.

Immunodeficiency

Glucocorticoids cause immunosuppression, decreasing the function and/or numbers of neutrophils, lymphocytes (including both B cells and T cells, monocytes, macrophages, and the anatomical barrier function of the skin. This suppression, if large enough, can cause manifestations of immunodeficiency, including T cell deficiency, humoral immune deficiency and neutropenia.

Main pathogens of concern in glucocorticoid-induced immunodeficiency:

Bacteria

- Enterobacteriaceae
- Legionella micdadei
- Listeria monocytogenes
- Mycobacterium tuberculosis
- Nontuberculous mycobacteria
- Nocardia asteroides
- Rhodococcus equi
- Salmonella species
- Staphylococcus aureus
- Streptococci.

Fungi

- Aspergillus
- Blastomyces
- Candida albicans and nonalbicans species
- Coccidioides immitis
- Cryptococcus neoformans
- Fusarium species
- Histoplasma capsulatum
- Penicillium marneffei
- Pseudallescheria boydii
- Zygomycosis.

Viruses

- Adenovirus
- Cytomegalovirus
- Herpes simplex virus
- Human papillomavirus
- Influenza/parainfluenza
- Respiratory syncytial virus
- Varicella zoster.

Other

- Cryptosporidiosis/lsospora belli
- Pneumocystis carinii
- Strongyloides stercoralis
- Toxoplasma gondii.

Withdrawal

In addition to the effects listed above, use of high-dose steroids for more than a week begins to produce suppression of the patient's adrenal glands because the exogenous glucocorticoids suppress hypothalamic corticotropin-releasing hormone and pituitary adrenocorticotropic hormone.

With prolonged suppression, the adrenal glands atrophy (physically shrink), and can take months to recover full function after discontinuation of the exogenous glucocorticoid. During this recovery time, the patient is vulnerable to adrenal insufficiency during times of stress, such as illness.

While suppressive dose and time for adrenal recovery vary widely, clinical guidelines have been devised to estimate potential adrenal suppression and recovery, to reduce risk to the patient.

The following is one example, but many variations exist or may be appropriate in individual circumstances.

- If patients have been receiving daily high doses for five days or less, they can be abruptly stopped (or reduced to physiologic replacement if patients are adrenal-deficient). Full adrenal recovery can be assumed to occur by a week afterward.
- If high doses were used for six to 10 days, reduce to replacement dose immediately and taper over four more days. Adrenal recovery can be assumed to occur within two to four weeks of completion of steroids.
- If high doses were used for 11–30 days, cut immediately to twice replacement, and then by 25% every four days. Stop entirely when dose is less than half of replacement. Full adrenal recovery should occur within one to three months of completion of withdrawal.
- If high doses were used more than 30 days, cut dose immediately to twice replacement, and reduce by 25% each week until replacement is reached. Then change to oral hydrocortisone or cortisone as a single morning dose, and gradually decrease by 2.5 mg each week. When the morning dose is less than replacement, the return of normal basal adrenal function may be documented by checking 0800 cortisol levels prior to the morning dose; stop drugs when 0800 cortisol is 10 ìg/dl. Predicting the time to full adrenal recovery after prolonged suppressive exogenous steroids is difficult; some people may take nearly a year.
- Flare-up of the underlying condition for which steroids are given may require a more gradual taper than outlined above.

Chemical Synthesis

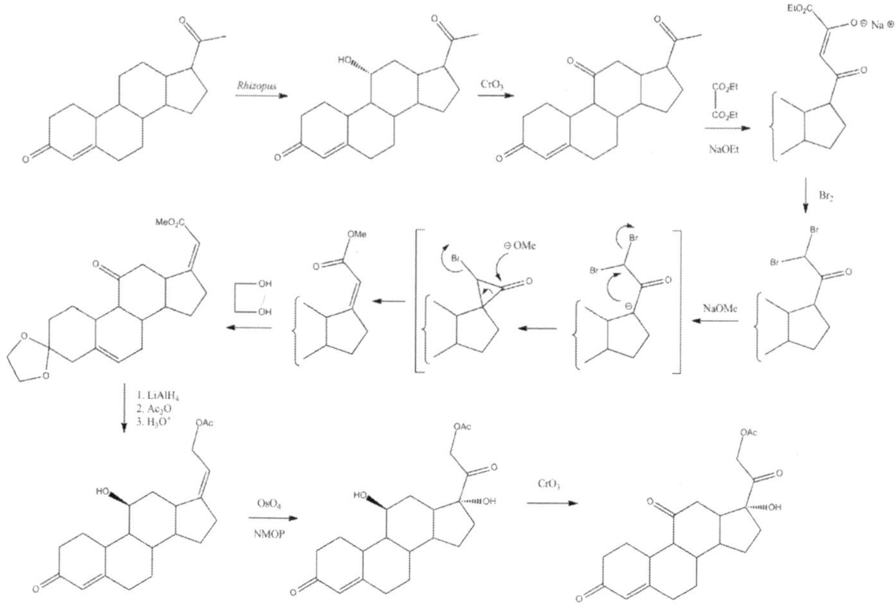

Evolution of Glucocorticoid and Mineralocorticoid Responses: Go Fish

In the premolecular biology era, before the cloning of the glucocorticoid receptor (GR) and mineralocorticoid receptor (MR), more was known about the binding of aldosterone, cortisol, corticosterone, and other corticosteroids to the GR than was known about their binding to the MR. The GR had advantages of being both present in reasonable quantities and sufficiently stable in tissue homogenates for steroid competition analyses. Moreover, establishment of the rat hepatoma tissue culture (HTC) cell line in Tomkins' laboratory provided the opportunity for parallel studies of the binding of cortisol, corticosterone, and other corticosteroids to the GR and their effects on GR-mediated induction of tyrosine amino transferase. In addition, HTC cell extracts did not contain corticosteroid binding globulin (CBG), which was present in tissue homogenates, and skewed measurements of the binding of endogenous glucocorticoids to the GR.

In contrast, the MR was much more difficult to study due to low levels of MR in the kidney, the main target organ for aldosterone, and MR instability in kidney homogenates. Moreover, CBG in kidney homogenates made it difficult to study the competitive binding of cortisol corticosterone, deoxycorticosterone, and other glucocorticoids to the MR. With effort, CBG was removed from kidney homogenates, and there was a surprise: aldosterone, corticosterone, and cortisol have similar affinity for the MR. Even more surprising was the demonstration that hippocampus, a nontraditional aldosterone target tissue, contained a MR with specificity similar to that in the kidney MR.

The cloning of the human MR by Arriza *et al.* established that aldosterone, corticosterone, and cortisol also had similar affinity for the human MR, although in the cell cultures studied, aldosterone was active at a 10-fold lower concentration than cortisol in stimulating MR-mediated transcription. The presence of mRNA for MR was verified in brain and heart, as well as traditional tissues, kidney and colon.

These studies raised important questions about aldosterone action. First, because free levels of glucocorticoids are 100-fold higher in blood than those of aldosterone, how could aldosterone bind to MR and regulate the mineralocorticoid response? Second, what is the function of a MR in the brain and heart?

A year later, two papers resolved the first question and, in the process, added the mechanism of tissue-specific expression of enzymes as a partner with steroid receptors in regulating the actions of corticosteroids and other steroid hormones. MRs in the distal tubule of the kidney are not activated by cortisol due to coexpression of 11β-hydroxysteroid dehydrogenase, now called 11β-HSD-2, which oxidizes cortisol to cortisone, an inactive steroid. Aldosterone forms an 11,18 hemiacetal in solution and, thus, is not a substrate for 11β-HSD-2; it can thus occupy the MR and regulate electrolyte transport in the presence of 11β-HSD-2. However, this mechanism does not apply to hippocampus or heart muscle cells, which contain MR and lack 11β-HSD-2. Indeed, understanding what the MR is doing in nontraditional tissues has been a lively area of research in the last decade.

In this issue of *Endocrinology*, Greenwood *et al.* report the cloning from the cichlid fish of one MR and three GR isoforms, their response to aldosterone and cortisol, and their expression in various tissues, which both clarifies our understanding of the actions of the MR and GR and raises additional questions about the evolution of responses to corticosteroids. Their work illustrates Dobzhansky's epigram "Nothing in biology makes sense except in the light of evolution", which is even more relevant now than in 1973. Indeed, evolutionary analyses of complete genomes from bacteria and eukaryotes have determined when various genes and regulatory networks evolved, and how they diversified as different life forms evolved.

In this regard, on the time scale of 3.8 billion years for life on earth, the adrenal and sex steroid response is relatively young, as GR, MR, and receptors for androgens, estrogens, and progestins are not found in invertebrates such as the fruit fly and worm. These receptors appear to have evolved at the base of the vertebrate line, some time before the Cambrian explosion, which occurred 545 million years ago.

The earliest evidence for steroid receptors is in lamprey, a jawless fish, which contains an estrogen receptor, progesterone receptor, and corticosteroid receptor; however, their ligands and the responses that they mediate are not yet known.

Steroid binding studies of a partial MR sequence and full GR from trout have provided some information on corticosteroid receptors in fish. However, with the newly cloned full-length MR and three GRs from cichlid, we have the necessary tools to study transcriptional activity of different corticosteroids and the functions of the MR and GR in fish. Common features between mammalian and fish receptors are likely to predate the divergence of fish and land vertebrates, which moves our understanding corticosteroid actions back about 450 million years, when mammals and fish last shared a common ancestor.

First, Greenwood *et al.* find that aldosterone and cortisol have similar EC_{50} values of 0.02 nm and 0.05 nm, respectively, for activation of cichlid MR. In parallel studies with rat MR, aldosterone and cortisol have EC_{50} values of 0.04 nm and 0.3 nm, respectively, indicating less selectivity in fish for these two corticosteroids. As found in mammals, cichlid MR is highly expressed in brain and heart, which supports an important function for the MR in these tissues. Unexpectedly, cichlid MR is not well expressed in kidney. Indeed, the levels in kidney are slightly below that in liver and spleen. Substantially more MR is expressed in the gill, which is a site for chloride transport in fish.

Cortisol has an EC_{50} of about 3–5 nm for all three GRs; aldosterone has about 100-fold lower potency. GR1 and GR2a have a DNA binding domain that is similar to other GRs. GR2b has an insert of 9 amino acids between each zinc finger domain that was first found in trout by Ducouret *et al.* and is also in founder GR. A recent search of

GenBank did not find this segment in other steroid receptors. Expression patterns of the three GRs are complex and will require additional research to understand. GR2 is more highly expressed than GR1 in brain, heart, gill, kidney, liver, and spleen. GR2b is the main GR2 receptor in gill and liver; GR2a is more highly expressed than GR2b in kidney and spleen; GR2a and GR2b are about equally expressed in brain and heart.

The report of Greenwood *et al.* needs to be considered with that of Kusakabe *et al.*, who cloned trout 11β-HSD-2, which regulates two different endocrine responses. As in mammals, trout 11β-HSD-2 oxidizes cortisol to cortisone, controlling access of active glucocorticoids to fish MR and GRs. However, 11β-HSD-2 also catalyzes oxidation of 11â-testosterone to 11-keto-testosterone, the active androgen in fish. Thus, expression of 11â-HSD-2 can both increase 11-keto-testosterone and decrease cortisol levels.

Due to its multiple biological activities, trout 11β-HSD-2 is highly expressed in kidney, gill, intestine, testis, ovary, heart, pituitary, liver, and skin. However, 11β-HSD-2 is very weakly expressed in brain; Kusakabe *et al.* needed 8-fold more brain mRNA than was used for other tissues to see a weak signal in a Northern analysis. If other fish also have low expression of 11β-HSD-2 in brain, then cortisol will occupy the MR and GRs in fish brain.

The high affinity of aldosterone for the cichlid MR is relevant to the still unresolved question: is aldosterone a fish steroid? This was a contentious issue even in 1967; most attempts to find aldosterone in fish have been unsuccessful. In some fish, trace levels of aldosterone have been reported. However, given that Greenwood *et al.* find that aldosterone has an EC_{50} of.03 nm for the MR, aldosterone could be biologically active at very low concentrations.

From an evolutionary perspective, aldosterone should not be the ancestral mineralocorticoid because synthesis of aldosterone is more complex than that of deoxycorticosterone and corticosterone, which have just as high an affinity as aldosterone for the mammalian MR. Parsimony suggests that corticosterone (or possibly deoxycorticosterone) was the ancestral ligand for the MR, and possibly for one or more fish GRs. Moreover, our search of the recently completed Fugu genome with human CYP11B2, aldosterone synthase, did not find a protein with greater than 45% sequence identity. This suggests that Fugu does not contain an aldosterone synthase, although it is possible that the Fugu genome is incomplete. Also, CYP11B2 may have been lost in Fugu but may be present in other fish.

Lastly, the complexity of corticosteroid action in cichlid uncovered by Greenwood *et al.* makes a strong case for studying the role of serum proteins in steroid specificity for the MR and GR in fish, in particular, CBG, for which we do not find an ortholog in the Fugu genome. In the last decade, CBG has been overshadowed by the importance of 11β-HSD-2 in regulating corticosteroid action. However, as was shown 30 yr ago, CBG regulates access of deoxycorticosterone, cortisol, and corticosterone to MR and GR, with aldosterone escaping this protein, just as aldosterone escapes 11β-HSD-2. Aldosterone may have evolved along with a CBG that sequestered corticosterone and other glucocorticoids. Studies of MR, GR, CBG, and aldosterone synthase in lobe-finned fish, such as lungfish and coelacanths, which belong to the fish class that is the ancestor of land vertebrates, should elucidate molecular events leading to the emergence of aldosterone as a mineralocorticoid in land animals.

As we celebrate the 50th anniversary of the discovery of aldosterone, it is comforting to know that its full story has not been told. More exciting discoveries, with important clinical applications, await the elucidation of the origins of corticosteroid action.

Mineralocorticoid

Mineralocorticoids are a class of steroid hormones characterised by their similarity to aldosterone and their influence on salt and water balances.

Physiology

The name mineralocorticoid derives from early observations that these hormones were involved in the retention of sodium, a mineral. The primary endogenous mineralocorticoid is aldosterone, although a number of other endogenous hormones (including progesterone and deoxycorticosterone) have mineralocorticoid function.

Aldosterone acts on the kidneys to provide active reabsorption of sodium and an associated passive reabsorption of water, as well as the active secretion of potassium in the principal cells of the cortical collecting tubule and active secretion of protons via proton ATPases in the lumenal membrane of the intercalated cells of the collecting tubule. This in turn results in an increase of blood pressure and blood volume.

Aldosterone is produced in the cortex of the adrenal gland and its secretion is mediated principally by angiotensin II but also by adrenocorticotrophic hormone (ACTH) and local potassium levels.

Mode of Action

The effects of mineralocorticoids are mediated by slow genomic mechanisms through nuclear receptors as well as by fast nongenomic mechanisms through membrane-associated receptors and signalling cascades.

Steroidogenesis, showing mineralocorticoids in ellipse at top right. Note that it is not a strictly bounded group, but a continuum of structures with increasing mineralocorticoid effect, with the primary example aldosterone at top.

Genomic Mechanisms

Mineralocorticoids bind to the cytosolic mineralocorticoid receptor. This type of receptor gets activated upon ligand binding. After a hormone binds to the corresponding receptor, the newly formed receptor-ligand complex translocates itself into the cell nucleus, where it binds to many hormone response elements (HRE) in the promoter region of the target genes in the DNA.

The opposite mechanism is called transrepression. The hormone receptor without ligand binding interacts with heat shock proteins and prevents the transcription of targeted genes.

Aldosterone and cortisol (a glucosteroid) have similar affinity for the mineralocorticoid receptor; however, glucocorticoids circulate at roughly 100 times the level of mineralocorticoids. An enzyme exists in mineralocorticoid target tissues to prevent overstimulation by glucocorticoids. This enzyme, 11-beta hydroxysteroid dehydrogenase type II (Protein:HSD11B2), catalyzes the deactivation of glucocorticoids to 11-dehydro metabolites. Licorice is known to be an inhibitor of this enzyme and chronic consumption can result in a condition known as pseudohyperaldosteronism.

18 Hydroxy 11 Deoxycorticosterone

18 hydroxy 11 deoxycorticosterone (also designated 18OH-DOC) is a steroid hormone probably used to conserve sodium and stimulate

hydrogen ion (or acid) excretion. 18OH-DOC lowers urine pH but has no effect on potassium excretion. This would seem to indicate that 18OH-DOC's primary purpose is to stimulate hydrogen ion or ammonium excretion.

Under low sodium intake 18 OH DOC is increased in serum. There is a marked increase in serum 18OH DOC after injection of insulin and this may be due to the hypokalemic (low serum potassium) tendency after a rise in insulin which in turn would make the serum more acidic. Since 18OH-DOC lowers urine pH (increases acidity) but has no effect on potassium excretion, this would seem to indicate that 18OH-DOC's primary purpose is to stimulate hydrogen ion or ammonium excretion.

Its use by the body to conserve potassium would be indirect by virtue of hydrogen ion's interference with potassium excretion. This interference is further indicated because injecting sodium bicarbonate or even hyperventilating (breathing rapidly beyond need) can triple potassium excretion. The daily rhythm for potassium and hydrogen ion excretion show a rather close inverse relationship, which gives additional circumstantial support to the supposition that they compete at a common site. 18OH-DOC is strongly dependent on the potassium cell or plasma content, because in potassium-deficient rats markedly less 18OH-DOC is converted to 18OH-corticosterone and less yet if sodium is deficient.

ACTH (a peptide hormone) has a large effect on 18OH DOC, causing 18OH DOC to go down to zero when ACTH does. This could be for the primary purpose of keeping serum immune enzymes and cell fluids at a high pH (alkaline) during internal infection, but not doing so during the intestinal infection of diarrhea, during which disease the resulting dehydration forces ACTH to decline. It probably is important normally to keep the vacuoles where pathogens are digested at a high pH because if the pH or alkalinity is not high enough, the pathogens inside the immune cells are not digested and thus released intact. So when an intestinal disease is not calling for ACTH to decline, the indirect potassium conserving attribute of 18OH-DOC by virtue of stimulating acid excretion would be valuable, as would increased acid excretion during internal disease.

18OH DOC may act primarily by blocking aldosterone's effect on potassium, and must have aldosterone to assist it with sodium. Nichols, et al., have been able to show that injection of 18OH-DOC, which raised blood levels of this hormone ten times, were more retentive of sodium than a similar amount of aldosterone. So there must be a synergism involved. At the same time, the ratio of sodium to potassium excretion declined very little for 18OH-DOC, while for aldosterone, the ratio fell to as little as 1/3 that of control men. This implies a considerable sparing of potassium by 18OH-DOC. Urine potassium excretion is not altered by 18OH-DOC injection.

Angiotensin II has very little effect on 18OH-DOC and is ambiguous (nor does serum potassium above 4.8 mEq/litre (187 mg)). This last is not surprising since 18OH-DOC should not be used by the body at high serum potassium. Under low sodium intake, 18OH-DOC rises in the serum. ACTH causes a marked increase in 18OH-DOC, probably by a generalised affect on the zona fasciculata of the adrenal cortex where 18OH-DOC is synthesized. So when it is necessary for sodium to be unloaded during the dehydration induced decline of ACTH during diarrhea in order to preserve osmotic pressure, the resulting 18OH-DOC decline would assist in this.

18OH-DOC is deeply involved in one of the three forms (at least) of hypertension (high blood pressure).

Pathophysiology

Hyperaldosteronism (the syndrome caused by elevated aldosterone) generally results from adrenal cancers. The two main resulting problems:

1. Hypertension and edema due to excessive Na+ and water retention.
2. Accelerated excretion of potassium ions (K+). With extreme K+ loss there is muscle weakness and eventually paralysis.

Underproduction, or hypoaldosteronism, leads to the salt-wasting state associated with Addison's disease, although classical congenital adrenal hyperplasia and other disease states may also cause this situation.

Pharmacology

An example of a synthetic mineralocorticoid is fludrocortisone (Florinef). Important mineralocorticoid inhibitors are spironolactone and eplerenone.

Hormone-Induced Spawning of Fish

The demand for fish for food, recreation, and ornamental aquariums is steadily increasing. Natural fish populations have declined during the last several decades because of environmental degradation and over-fishing.

This has resulted in an increased effort in the development of techniques for hatchery production of fish. Traditional aquaculture species such as trout, catfish, common carp, golden shiner, and goldfish reach sexual maturity and spawn in hatcheries or ponds, when conditions are appropriate.

However, a number of fish species that have or potentially have great economic significance for aquaculture do not reproduce spontaneously in captivity. Many of these fish spawn in environments that are nearly impossible to simulate in a hatchery.

Hormone-induced spawning is the only reliable method to induce reproduction in these fishes. Hormone-induced spawning of fish has been used for almost 60 years.

Surprisingly, the same procedures, with only minor modifications, have been used to spawn an entire range of fishes from the ancient sturgeon and paddlefish to carp, catfish, salmon, sea bass, redfish, snook, and mullet. In addition to breeding other desirable fish species, induced spawning can be used to:

- produce hybrids that are different from the parent species;
- produce sterile polyploid fish (for example, sterile triploid grass carp for aquatic weed control);
- synchronize reproduction of large numbers of fish for simultaneous spawning, thereby simplifying production and marketing of the fish;
- produce fry outside the normal spawning season for maximum hatchery production and to provide fish when the price and market demand is greatest; and
- maximize survival of fry under controlled hatchery conditions.

Proper Fish Handling: The physical injury and physiological stress of capturing, handling, transporting, injecting, and holding brood fish can have a greater detrimental effect on spawning success than almost any other factor. Fish must be handled carefully and optimum water conditions must be maintained to minimize stress.

The importance of proper handling and water quality cannot be overemphasized. Female brood fish ready for spawning are in a particularly delicate condition. When female fish are stressed or injured, they may undergo rapid physiological changes that can result in the breakdown (resorption) of eggs in the ovary.

Crowding, dissolved oxygen depletion, rapid changes in temperature, and osmotic imbalance are well known causes of stress and must be avoided. Suboptimum conditions, while not immediately lethal, may stress brood fish, resulting in delayed mortality or failure to spawn.

Figure: Golden shiner

Reducing stress and injury to brood fish can greatly increase the success of hormone induced spawning.

Determine Sexual Maturity: The external appearance of brood fish has long been used to assess the stage of sexual development. In some species, males change in appearance during the spawning season. These physical changes make it relatively easy to identify sexually mature males.

However, secondary female sex characteristics such as plumpness of the abdomen and redness of the vent are extremely subjective and can be misleading. Sampling the eggs and sperm of the brood fish eliminates the guesswork in determining the stage of sexual development. Milt can usually be stripped from males of most species when they are ready for spawning by applying gentle pressure to the abdomen between the pelvic fins and the vent. Sperm viability usually can be determined by observing motility with a microscope.

Several methods are available to determine the developmental stage of the eggs in the fish's ovary. The diametre and appearance of the egg and the position of the nucleus in the egg are visual indicators of development. The steroid assay procedure determines the physiological response of the eggs to hormones. Both require that an egg sample be collected.

The ovary can be sampled with either a rigid or flexible tube (catheter). An egg sample can also be taken by making a small incision along the belly or side of the fish. This technique is commonly used for sturgeon and paddlefish.

An understanding of sperm viability and egg stage development will greatly improve the success of hormone-induced spawning of fish.

Control of Reproduction in Spawning Fish

Reproduction in fishes is regulated by external environmental factors that trigger internal mechanisms.

The final event of the reproductive cycle, the release of eggs and sperm resulting in spawning, can be controlled by either placing the fish in an appropriate environment or by changing the fish's internal regulating factors with injected hormones or other substances.

The internal mechanisms that regulate spawning are similar for most fishes. The external environmental factors that control reproduction, however, vary considerably among species.

For this reason, more is known about the internal regulatory mechanism of fish reproduction than the specific environmental requirements for spawning each species.

Environmental factors that have been shown to play a significant role in the reproductive cycle are:

- photoperiod
- water temperature
- water quality (e.g., dissolved oxygen, pH, hardness, salinity, alkalinity);
- flooding and water current
- tides and cycles of the moon
- weather cycles (e.g., atmospheric pressure, rainfall)
- spawning substrate (e.g., aquatic plants, sticks, gravel, spawning caverns)
- nutrition

- disease and parasites
- presence of other fish

These factors do not function independently of each other, but are interrelated. The internal mechanism that regulates the process of reproduction in fish is the brain hypothalamuspituitary- gonad chain.

Hormone-induced spawning techniques influence this sequential mechanism at several levels, by either promoting or inhibiting the process. The primary substances used for hormone- induced spawning have been: (1) pituitary extracts and (2) purified gonadotropin to stimulate the ovaries and testes; (3) LHRH analogs (LHRHa) alone or in combination with (4) dopamine blockers which enhance the potency of LHRHa to stimulate the pituitary; or (5) steroids to stimulate the gametes directly.

The appropriate hormone preparation should be selected on the basis of the species to be spawned and the availability y of the hormones.

Preparation of Hormones: The hormones must be mixed and stored properly to prevent contamination and preserve potency. The proper dosage must be calculated for the brood fish, and the optimum injection schedule must be used for best results. To calculate the proper dosage, (1) the recommended dose, (2) approximate weight of the brood fish, and (3) the desired volume of the injection must be determined.

The quantity of hormone to be injected can then be calculated from the weight of each individual brood fish and the appropriate injection schedule.

Taking the Spawn: The eggs and milt of fish can be taken by several different methods: (1) tank spawning; (2) hand stripping, and (3) surgically removing the eggs. The method of choice depends on the fish species, hatchery facilities, experience and skill of the hatchery staff, and the desired manipulations of eggs, sperm, or fertilized eggs.

Tank spawning is the simplest method for obtaining a hormoneinduced spawn. Brood fish of both sexes are placed together in the spawning tank following injection(s). The female ovulates when she is physiologically ready. The males stimulate the female to release the eggs and fertilize the spawn.

Hand stripping is commonly used for taking the spawn of many species of fish. Brood fish are separated by sex prior to hormone injection to prevent spawning in the holding tank. It is important to determine the exact time of ovulation when hand stripping.

In many species, egg quality can deteriorate rapidly if the eggs are not taken shortly after ovulation. For most species, ovulation can best be verified by checking the female to determine when eggs flow freely from the vent. To strip the eggs, the fish is held slightly on her side, tail down; gentle hand pressure is applied to the abdomen, moving toward the vent. The stream of eggs is directed into a clean, dry bowl positioned so that water from the fish does not drip onto the eggs.

It is important to insure that no water comes in contact with the eggs until after the milt is added and mixed. Water activates the sperm and also causes the opening through which the sperm enters the egg to close. Because the internal anatomy of fish vary greatly, hand stripping may be difficult in some species.

Sturgeon and paddlefish have no ovarian sac; the eggs are released into the abdominal cavity during ovulation. The best method for taking the spawn of these species is to surgically remove the eggs. For delicate species that seldom survive the rigors of hand stripping, humanely killing them and surgically removing the eggs may be the best option. In addition, more eggs can usually be obtained by this method than by hand stripping.

Fertilizing the Spawn: The eggs obtained by hand stripping or surgical removal are usually fertilized with fresh milt. Males are captured, wiped off, and held belly down over the bowl containing the eggs. The portion of the abdomen posterior to the pelvic fins is gently massaged to extrude the milt onto the eggs.

Milt can be collected from males and stored up to three weeks prior to stripping eggs. Ovulated eggs of many species such as white bass, sturgeon, paddlefish, and common carp become sticky after water is added. During natural spawning, this stickiness causes the eggs to become attached to rocks, sticks, or aquatic plants. Catfish eggs are connected by a sticky matrix that holds the eggs together in a mass.

In the hatchery, this stickiness causes problems during incubation. Siltclay, Fuller's earth, or bentonite suspension, urea and salt solution, and tannic acid solution are preparations commonly used to deactivate the sticky layer of fish eggs. In addition, the gelatinous matrix of catfish egg masses can be dissolved with sodium sulfite so the eggs can be incubated in hatching jars.

Induced hatchery spawning of fish requires a continuous series of decisions, any of which if improperly made, can diminish or completely obliterate the success of the project. There are many ways to fail at each step and only a very few that are productive. Consistent performance requires strict attention to detail.

Captive Production of Ornamental Marine Fishes

As with the development of our species in going from being hunter-gathers of foods to their controlled growth and reproduction, marine aquaculturists have begun to make inroads in successfully breeding and rearing of marine fishes. Indeed, a few species (the Clownfishes, subfamily Amphiprionae and Neon/Cleaner Gobies, genus Gobiosoma) are largely produced in captivity nowadays, doing more than reducing correspondingly their wild-collection. Captive-produced marines, as with freshwater have proven hardier, more adaptable to aquarium conditions. They much more readily take prepared foods and are often specific pathogen free.

Though the vast majority (likely more than 99%) of marine fishes used for ornament are still wild-collected, there are great improvements in capture of healthy broodstock, knowledge of modes of their reproduction, induced and natural spawning and breeding, culture of useful foods, and overall husbandry techniques which foretell the possibility of captive production of any and all species.

Fishes are induced to spawn by altering environmental parameters, particularly temperature and photoperiod, and hormonal manipulation. These variables are generally constant for a given species, but vary widely amongst cultured fishes.

Spawning, Breeding: Some fifty-four families of fishes include livebearing species that utilise internal fertilization with either no further nourishment from females (ovoviviparity) or placenta-like structures between their young and their bodies (viviparity). Forty of these fish families are cartilaginous fishes (sharks and rays) and the remaining marine ones of little notice (currently to marine aquarists). Additionally there are fishes that engage in internal fertilization, placing their sticky embryos at a later time. None of these species are either presently utilised by aquarists. Hence, we are almost exclusively concerned with spawning species, those that broadcast their sex cells into the environment when considering the captive production of ornamental marine fishes.

History: Early records at trials at culture of ornamental marine fishes likely don't exist. Many "casual aquarists" of the sixties experimented with home and some commercial production of Clownfishes. Their accounts are anecdotal at best. Friese (1971) reminisces over hobbyist accounts of partial successes of breeding, rearing marine fishes (seahorses, pipefishes, some gobiids, a few pomacentrids) in the 1960's, and cites "recent" work to that time on multigenerational breeding done at the Wilhelma Aquarium in Stuttgart, Germany by Dr. W. Neugebauer of *Hippocampus, Dunckerocampus* and *Amphiprion* spp.

Hoff (1985) relates the early beginnings of commercial ornamental marine fish culture in the United States with his involvement with Instant Ocean Hatcheries (1974-1985), Aqualife Research (moved to Walker's Caye in the Bahamas from Florida, founded by Martin Moe in 1972, who is still engaged in active research) and Sea World (San Diego, run by Chris Turk) who have all ceased production.

These early ventures likely failed commercially as Hoff states for Instant Ocean Hatcheries, due to too much overhead cost. They all were able to sell all (mainly Clownfishes) they could produce.

Current commercial facilities in operation include Tropic Marine Centre in the U.K., ORA (Oceans Reefs Aquariums) in Florida in the U.S.A, and C-Quest in Puerto Rico, Mangrove Tropicals (clowns and Dottybacks) and Ocean Rider (seahorses), both of Hawaii and South Australian Seahorse Marine Services (*Hippocampus abdominalis*). Experimental/university labs that are engaged in research involving marine pet-fish species include The Aquarium Complex/Marine & Aquaculture Research Facilities Unit (MARFU) at James Cook University in Queensland, Australia, Guam Aquaculture Development & Training Centre, Hofstra University Marine Laboratory in Hempstead, New York, various Sea Grant locations in the United States (principally in Hawaii),

At Right, Some of ORA's Captive-Produced Clownfishes.

Amphiprion ocellaris Cuvier 1830, the "False" Percula or Clown Anemonefish, or Ocellaris Clown. Indo-West Pacific; eastern Indian Ocean to Australia, to Philippines, to southern Japan. To a little over four inches maximum length. Bred in captivity including beautiful northwest Regular and Australian dark variety mid-juvenile and adult below. Orange overall (except for melanistic forms), with three broad continuous body bars with narrow black margins (vs. thick ones in A. percula).

Amphiprion percula (Lacepede 1802), the "True" Percula or Orange Clownfish. Western Pacific; New Guinea, GBR, Solomon Islands, Melanesia. To about four inches in length. Mutualistic with *Stoichactis* and *Radianthus* anemones.

Amphiprion melanopus Bleeker 1852, the Red and Black Anemonefish, Melanopus Clown. Variable amounts of red and black, with or w/o a "Tomato Clown" like single white head bar. Indonesia, to the Society and Marshall Islands. To nearly five inches in length.

Amphiprion ephippium (Bloch 1790), the Red Saddle Anemonefish. Eastern Indian Ocean; Andaman and Nicobar Islands, Thailand, Malaysia, Java and Sumatra in Indonesia. To five and a half inches maximum length. This three inch female in the Singapore Aquarium. Body, fins orange to reddish.

Amphiprion perideraion Bleeker 1855, the Pink Anemonefish. Indo-Pacific. To four inches in length. Narrow midline white stripe and head barring. This one in a *Heteractis magnifica* in Western Australia.

Amphiprion polymnus (Linnaeus 1758), the Saddleback Clownfish. Western Pacific; Ryukyu Islands to the Gulf of Thailand, Indonesia. Northern Australia and Solomon Islands. To five inches in length. Distinctive saddle-like marking on mid-body. Below: a pair in a typical silty/sandy setting in a Stichodactyla haddoni anemone (also inhabits Heteractis crispa in the wild), off Pulau Redang, Malaysia, and in an aquarium. A darker one from N. Sulawesi tending its eggs.

Premnas biaculeatus (Bloch 1790), the Spine-Cheek Anemonefish, Maroon Clownfish. Bright red to orange body and fins, with three narrow white body bars and prominent opercular spine. Females to six inches, males about half that size.

Gobies, Family Gobiidae Though many species of gobies of this large family have been captively spawned and reared, only the Cleaner Gobies, principally of the genus Gobiosoma are presently commercially produced. Along with the Clownfishes the cleaner gobies make up the bulk of the ornamental marine fish culture market.

Gobiosoma oceanops (Jordan 1904), THE Neon Goby. Tropical West Atlantic; southern Florida to Belize. To two inches in length.

Elacatinus puncticulatus (Ginsburg 1938), the Red Headed Neon Goby. Eastern Central Pacific; Sea of Cortez to Ecuador. To under two inches in length. A real beauty and tough, but small addition to reef tanks. These ones in captivity at a wholesaler and retailers.

The Sharknose Goby, *Gobiosoma evelynae* Bohlke & Robins 1968. Tropical West Atlantic; Bahamas to Venezuela. To about two inches in length. Variable in colour. *Gobiosoma evelynae*; blue and white striped adults in the Bahamas.

Gobiosoma randalli Bohlke & Robins 1968. Tropical West Atlantic; Puerto Rico to Venezuela. To under two inches in length. Here is an individual in captivity.

Dottybacks, Family Pseudochromidae: Dottybacks can be very aggressive when wild-collected. As with many cultured freshwater fishes, captive-produced generations are much more easygoing.

Pseudochromis fridmani Klausewitz 1968, the Orchid Dottyback. Known only from the Red Sea, but cultured in commercial numbers. To three inches in length. This photo taken in Sharm, Red Sea.

Pseudochromis sankeyi Lubbock 1975, the Striped Dottyback. Gulf of Aden and lower part of Red Sea in the northwestern Indian Ocean. To three inches in length. Aquarium photo.

Pseudochromis flavivertix, the Sunrise Dottyback. Yet another pseudochromid out of the Red Sea (and Gulf of Aden to the south).

Pseudochromis splendens, the Splendid Dottyback. Eastern Indonesia and Northwest Australian waters. To three inches in length. A commonly cultured, and for the group, very peaceful Dottyback species. This sunken bellied specimen photographed in a LFS in New Jersey.

Pseudochromis aldabraensis Bauchot-Bautin 1958, the Orange Dottyback. Indian Ocean, Arabian Gulf, down to Aldabra, over to Sri Lanka. To four inches in length. Shy, but a great aquarium beauty when kept with suitably larger fish specimens. Aquarium image.

Pseudochromis steenei Gill & Randall 1992, the Lyretail Dottyback. West Pacific. To five inches in length. Amongst mean fishes like the Dottybacks, this species is a pure terror. Absolutely to not be trusted with any other fishes of small size or gentle natures. Aquarium image.

Cardinalfishes, Family Apogonidae: A few species of this male-mouthbrooding family have been spawned and reared in captivity. As of now, the only commercially valuable species under culture is the Banggai Cardinal.

Pterapogon kauderni Koumans 1933, the Banggai Cardinalfish. Restricted in distribution to Banggai Island, Indonesia, though commercially produced in good numbers in Indonesia and elsewhere. To three inches in length. A darling of the ornamental aquatics industry and hobby. Readily reproduced in captivity. Young cluster about the spines of the Urchin *Diadema setosum* when threatened as pictured.

Seahorses and Pipefishes, Family Syngnathidae:These ever-popular aquarium fishes have dismal survival histories coming from the wild that are entirely reversed with captive-produced stocks. TMC of the UK has "closed the loop" producing successive generations of some species of pipes, and Ocean Rider of Hawaii is enjoying great success with production of a few seahorse species. Institutions like public aquariums have had success with breeding and rearing more than a dozen species of horses to date.

Doryrhamphus dactyliophorus (Bleeker 1853), Ringed Pipefish. Indo-Pacific; Red Sea, east Africa to Samoa. To 7 3/4" in length. A common offering in the pet-fish interest. Do buy tank-bred specimens. Wild ones fare poorly. Aquarium and N. Sulawesi images.

Hippocampus abdominalis Lesson 1827, the Big-Belly Seahorse. Southwest Pacific; Australia and New Zealand. To a foot in length. Now cultured in good numbers by OMLAS Pty Ltd (Seahorse Australia) in Tasmania for the aquarium trade.

Hippocampus histrix Kaup 1856, the Thorny Seahorse. Indo-Pacific; East Africa to Hawai'i. To 17 cm. in length. N. Sulawesi photo. Common species in the pet-fish and Chinese medicine interests. N. Sulawesi pix.

Hippocampus kuda Bleeker 1852, the Common Seahorse. Indo-Pacific; Pakistan, India, to Hawai'i, Society Islands. To a foot in length (stretched out). Found in calm waters amongst algae, seagrass. N. Sulawesi image at right, captive ones below.

You are referred to the bibliography below and *The Breeders Registry* in print or on the Internet especially for a review of the compendia on particulars of raising the young of the several syngnathid species that have gone through generations in captivity. Such a daunting task involves even more aquariums and culture of microalgae and small zooplankters.

Seahorse Culture Tanks, Birch Aq., San Diego

Production Considerations: Several critical elements must come together in aquaculturing captive ornamental fishes. Selection of suitable species, securing of broodstock, conditioning, the actual physiology of gamete production and release through environmental and/or hormonal manipulation, provision of grow-out facilities and supplying of appropriate foods.

Species Selection: Criteria for selecting potential species for captive propagation include compatibility, colour and markings, size, inherent hardiness to captive conditions/shipping and handling, and interesting behaviour.

Compatibility is key as there is a huge range of desirable behaviour amongst reef fishes. Some are absolute terrors that even with successive generations of captive breeding are proving troublesome. The best example here is *Pseudochromis steenei*, the Lyreback Dottyback. This fish is likely responsible for many aquarists outright quitting the hobby due to its predaceous nature.

Broodstock: Many breeding attempts are made impractical if not impossible by the selection of inopportune broodstock. Sufficient numbers of potential spawners or breeders must need be secured to afford the luxury of their controlled upbringing and possible sacrifice to determine viability.

Conditioning: Keeping broodstock in optimized, stable environments is absolutely key to captive breeding programs. As such culture facilities do their utmost to properly house their breeders. Most are kept in recirculated, centralised filtered tanks, with automated controls for water quality, including dosers for values like pH, probes for temperature, and regular testing for metabolites like nitrate, phosphate.

Diet: Bogenschutz and Clemens (1967) were amongst the first to scientifically explore links between nutrition and gonadal development. They demonstrated gonadal regression induced by restricted diets, accompanied by an inversion of the basophil/acidophil ratio of the mesoadenohypophysis and a reduction of gonadotropin content. These conditions were reversed with adequate food. These authors stressed the interplay between photoperiod and diet and pointed out that optimum benefit from photoperiod manipulation can be overridden by poor diets.

Often special foods with vitamins, HUFAs (Highly Unsaturated Fatty Acids), and minerals added are specifically made, tailored to the individual species. Some excellent businesses like England's Tropic Marine Centre use this opportunity to tie in sales of their proprietary frozen food line (Gamma Foods). Use of dried foods and relatively hollow nutrient content foods like adult brine shrimp is discouraged for conditioning breeders.

Disease Prevention: As part of a working plan of maintaining broodstock attention must be paid to preventing the introduction of disease-causing organisms. To mention them again, TMC (Tropic Marine Centre) should be cited for the exclusion of such pathogens. Through careful quarantine, use of antibody treatments, their entire facility is specific pathogen free.

Gametogenesis & Reproductive Behaviour: Marine fishes produce and release sex cells based on maturity of the individuals, their nutrition and overall health, triggered by cues from the environment (temperature, light/dark duration, tides, presence of conspecifics, mates'¦) that in turn influence their hormonal/endocrine systems. Along with endocrine control there is a steady, intimate, more sudden interplay of the fishes' nervous system, feeding in and coordinating activity through their eyes, hearing, lateralis system, senses of smell, and memory.

Conditioning and triggering of actual spawning involves combining knowledge of modes of reproduction, social factors such as sex ratios, environmental manipulation and possibly direct/exogenous hormonal administration.

Hormonal Manipulation as an Aquacultural Technique: All fish behaviours are hormonally mediated, with much the same hormones, pathways as higher vertebrates. Hormonal manipulation of heretofore "difficult" species to reproduce involves the injection of exogenous hormones (pituitary extracts, HCG/Human Chorionic Gonadotropin, PMSG/Pregnant Mare Serum Gonadotrophin among other preparations.) intracoelomically (into the body cavity), intramuscularly (into the muscle), and in rare cases intracranially (into the brain) as has been done with caviar-producing Sturgeons.

Though not commonly employed in captive pet-fish production, widespread use of hormonal manipulation has allowed the breeding of many aquatic source protein fishes for human consumption. Hormones are slow-acting chemical messengers. Along with the faster acting central nervous system the tissues producing hormones (the endocrine or "crying within" system) serve to moderate, direct and sustain the physiology of all animals and plants. Combined with environmental manipulation, exogenous hormone administration can be used to provide seed all year long rather than rely on wild-collection of seasonal broodstock or young.

Gonadotropins are a class of hormones that instigate the production, development and release of sex cells. Hypophysation involves the injection of gonadotropins from one animal into another to bring on the physiology and/or behaviour of reproduction. In practice with fishes, sometimes sexually-ready broodstock is sacrificed and their pituitaries (part of the brain that produces gonadotropins) utilised alone, or with mammalian-derived hormones or with these alone to cause changes in parental stock gametogenesis and sex-cell release.

Issues like standardization of dosage, determination of maturity of recipient fishes and nutritional requirements particularly need to be further explored for ornamental fish hypophysation. The first use of hormonal manipulation in fishes was in 1930 by B.A. Houssay of Argentina, who induced premature birth in viviparous fish by injecting pituitary glands recently removed from other fishes.

The author is aware of the use of hormonal manipulation being used experimentally on marine fishes for the ornamental trade, their use in marine food-fish production, and continued use in freshwater pet-fish species like Arowanas (Scleropages species) and Pangasiid Catfishes. However, as ongoing protocols, only environmental manipulation (principally temperature and photoperiod) are utilised to standardize, induce seed production and spawning. Bear in mind that what is being "manipulated" by changing environments are the broodstock's endocrine systems themselves. That these are variable and specific to species. The environment in all its elements (nutrition, social inputs, light, disease-causing organisms'¦) is the source of phenomena determining hormonal secretion.

Environmental Manipulation: The environment is the source of phenomena determining hormonal secretion. Several factors, photoperiod, temperature, metabolites, pheromones, light strength and temperature shock control release of gonadotropins, which in turn induce gamete production and concomitant reproductive behaviour. Reproduction may occur out of season by manipulation of temperature and photoperiod alone.

Studies such as Moyer et al. (1993), and compilations like Thresher's Reproduction in Reef Fishes (1985) point the way to how seasonality, particularly temperature and photoperiod influence natural reproduction. Additionally, social structure (inter- and conspecifically) plays a role in many species studied (e.g. Thalassoma spp., Warner and Hoffman (1980)). All fishes have discernible influences, cues that coincide with environmental changes'¦ typically tied in with the absence of predators of spawners and/or their young, availability of natural foodstuffs, reproduction of other life forms.

The reproductive behavioural plasticity of many important cultured species is demonstrated in Clownfish culture. These damselfishes are always found in intimate association with actinarian hosts (sea anemones) in the wild. In captive culture they are almost always kept and bred without these animals, and do reproduce very readily on provided hard substrate surfaces.

A further example of improved culture through just manipulation of environmental factors is offered: The Ayu (*Plecoglossus altivelis*) , a smelt-like fish cultured in Japan during the Summer months, normally spawns in Oct.- Nov., after which most of the fish die. Due to lower water temperatures that time of year, it's difficult to maintain live food organisms in adequate densities in rearing ponds to feed fry. An adjustment of photoperiod resulted in successful acceleration of sexual maturation (August), enabling rearing of fry when natural food was abundant. Spawning was also delayed to prolong the life of adults. The light period was extended to 18 hours per day with artificial light (Aug.- Oct.). Reproduction was retarded and adults were marketed out of season in Feb. (Kuronoma 1968).

As an interesting follow-up, recent trends in culture of the Ayu now include successful transfer of genetic material from rainbow trout. Trials have double their weight and increased length per time by a factor of about 1.3 (Cheng et al. 2002).

Temperature: The relative importance of temperature as an instigator of reproductive physiology and behaviour varies per species, investigator and experiment, but all agree a certain range and lowering or elevating optimizes results. Due to the nature of commercial enterprises, much of actual technique and applied values is of proprietary nature with ornamental marine fishes. Some unrelated species examples include Smiglieski (1975) who stated that temperature was the controlling factor in his work with flounders. Yamamoto et al. (1966) observed spermatogenesis in goldfish (Carassius auratus) below 14 C. and accelerated spermiation above 20 C.

Often, if much temperature fluctuation occurs reproductive behaviour will cease (Kaya 1973, Lofts et al. 1968). As most marine fishes are summer spawners, a steady increase in photoperiod (to 14-16 hours per day) and temperature (25-27 C., 77-81 F.) can be utilised to stimulate repeated spawning. For some fishes (e.g. Clownfishes) such manipulation can insure spawning every two or so weeks, for other species up to a three month cycle of condensing seasonal changes is required.

Haydock (1971) has observed temperature threshold below which gulf croaker will not hydrate or ovulate. For each species or sometimes race there are limits, both high and low, and rates of change in temperature which will eliminate reproduction physiologically and behaviourally.

Light: As with temperature, there are optimal amounts of strength, quality and duration of light in relation to reproductive events. On work with bluegill sunfishes it has been found that a longer photoperiod of 16L/8D vs. 8L/16D induced a better gonosomatic index (relative weight of gonad versus body; G.S.I. = ovary wt./total body wt. X 100). Males were more aggressive, dug more nests and fertilized more eggs.

Hoar (1969) sums up the effect of longer photoperiods by stating that they stimulate secretory activity of the anterior pituitary gland, inducing presexual behaviour. The pituitary Lutenizing Hormone activates interstitial tissue of the gonads which produce gonadal steroids which dominate sexual phases, taking complete control during parental phases (Kaya & Hassler 1972). Other investigators have shown that elevated temperature and prolonged photoperiod together increased gonad maturity in green sunfish, *Lepomis cyanellus.*

Other Factors: Photoperiod and temperature have received most attention and are generally considered to be of greatest importance in inducement of sexual maturity and spawning (Jhingran 1969). Other factors such as the effects of other environmental stimuli; meteorological (rain, floods, etc.) and water conditions (pH, ammonia, carbon dioxide, turbidity'¦), specific gravity (Walker and Herwig 1976), on controlled breeding of food and ornamental fishes has been investigated. In several species, the presence of con- and/or heterospecifics is important. In bluegills it has been found that when males are present, the most aggressive female has greater ovarian development. Another application of environmental control is employed in mullet culture in Israel. By keeping adult mullet in freshwater ponds, denying them access to the sea where they migrate to spawn, fish-culturists are able to extend the spawning season (Jerome 1975).

In summary, investigations into seasonal cues in the wild that bring on pre-spawning conditioning and spawning have proven fruitful for artificial culture of ornamental marines.

Criteria for Judging Sexual Readiness: Biological and Chemical Assay

Some species of ornamental fishes are raised in groups and maintained as breeding pairs (e.g. Clownfishes, Dottybacks, Gobiosoma gobies) or harems (dwarf angels of the genus Centropyge), while others are placed together only for the purpose of spawning. Several methods are applied towards determining sex and sexual maturity of spawners. For most of the current ornamental marine fishes under culture, formation of breeding pairs is done by maintaining numbers of individuals in larger (tens to hundreds of gallons) systems over months time, having them naturally "pair-up", then moving the pairs to small breeding tanks, generally moving them rather than their spawn and media it's attached to replicate containers.

In the use of hormonal manipulation it is often necessary to sacrifice some of the broodstock to assess their reproductive stage. Injection of hormones in an otherwise unripe adult will not generally induce gametogenesis or ripening of eggs if the breeders are otherwise unconditioned. Determination of spawning-readiness is sometimes associated with colour or marking changes, distension of the body. There are chemical assays of body fluids which can be used as guides of readiness, but these are not as commonly employed as much as simple hand-stripping of gametes, their mix and microscopic examination as a guide to broodstock fitness.

Spawners/Spawning

Spawning entails two strategies, either dispersement of eggs into the environment directly (pelagic) or placement of eggs (benthic) and directed spraying of spermatozoa to fertilize them. Almost all cultured species, indeed nearly all bony fishes have a pelagic larval developmental stage, a delicate time of days to weeks, sometimes months when they are moved about by ocean currents, hopefully avoiding predation and finding adequate forage.

Due to their greater ease of rearing, most cultured marine aquarium fishes are benthic spawners, with the tube-mouthed fishes (seahorses and pipefishes) and the mouthbrooding jawfishes and cardinalfishes being specialised cases. These fishes provide parental care, and though their spawning numbers are lower than broadcast spawners, their young hatch out in much more advanced stages of development, accepting larger food items and therefore enjoy higher survivability.

An example of Gobiosoma gobies: A few inches length of 1/2 inch PVC pipe is placed on the bottom of each pairs enclosure to serve as a spawning substrate. Eggs are adhered to the inside of the pipe by the female and fertilized by the male. Males guard the deposited, fertilized eggs until hatching, typically in 3-7 days, depending on species and temperature. Methods vary with Gobiosoma hatching protocols. Some culturists leave the egg mass with the male in attendance till the young are free-swimming, removing the male at that time. Others induce hatching by gently pipetting water over the egg mass when the embryos eyes become fully pigmented and the yolk sacs are no longer visible.

About Collection of Wild Spawn, Larval Fishes

If access is ready, there is a possibility of collecting wild produced larvae. This has been done for several species, including one of the two native pomacentrids of California, the Garibaldi, *Hypsypops rubicunda*, by the author. Guarded nests were combed of their developing eggs and young, transported in closed jars to the lab and grown out with varying degrees of incidental mortality. Unfortunately many young fish were found to be parasitized by intestinal nematodes.

Collection and rearing of post-settled larval ornamental marine fishes is a successful business in French Polynesia at present. Small fishes near to settling stage are caught in fine nets at night time on the reef and transported to land-based grow-out facilities. Wild-collected seed and larval sources bear the infectious and parasitic disease difficulties of larger wild-collected individuals, but have been shown to adapt with far greater facility to captive conditions.

Pelagic eggs and young can be collected with plankton nets (300-500 micron mesh) drawn slowly through the water or held stationary in currents. This is assuredly a "shot-gun" approach that leads to a very large, variable mix of species.

Rearing of Larvae, Young: Massive mortalities in the wild are the rule for initial few hours to days of fish larvae and young. For the majority of species ninety some percentiles are typical within the first few days, with gametes, embryos and fry being swept out to sea or being consumed by predators. Understandably, conditions are much more constant and less dangerous in captive culture. With provision of ready nutrition, an absence of predators, the majority of young per batch will hatch out and develop.

Culture of Foodstuffs: Toonen (2002) rightly points out that suitable food availability is key to successful aquaculture of marine animals. This is likely the largest source of mortality for cultured (and wild) reef fish larvae and fry. Having appropriate food organisms and prepared foods present almost continuously is required for optimal growth and survival.

Here again, there has been syncretization in wild studies such as Riley & Holt (1993) in studying gut contents of larval fishes as well as documented trials and errors in trying to grow and feed cultured foods (principally algae, rotifers, crustaceans, e.g. Hoff, 1999.) to cultured fish larvae. The road to discovering and providing proper, palatable foodstuffs at appropriate intervals and attractive formats has been a long and expensive one in many cases.

Plankton Culture

Various species of phyto and zooplankton are cultured to feed different species of larval fishes, often in a dual step fashion, growing phyto-plankton to feed the zooplankton to feed the fishes in turn. The plankton cultures are typically kept under controlled conditions (light, temperature) in separate rooms to prevent contamination. Some examples of commonly employed cultured live foods are detailed below.

Of the microalgae *Isochrysis galbana* can be grown under fluorescent lights per Hoff and Snell (1987). Isochrysis is used both as a water conditioner and as food for zooplankton. A common rotifer, *Brachionus plicatilis*, is often used as a first food, often followed by nauplii of the Brine Shrimp, *Artemia salina*. Rotifers can be cultured in as small as 10-gallon aquaria at a salinity of 25ppt, on a combination of Culture HUFA (tm) from Salt Creek, Inc., and concentrated *Isochrysis* paste from Reed Aquaculture, inc. Culture densities typically range from 100-250/ml.

Certainly key in fish culture worldwide is the hypersaline crustacean called Brine Shrimp, *Artemia salina*. Rather than having to collect and hold them as food items, *Artemia* cysts are often decapsulated with household chlorine bleach that then can be refrigerated in a saturated salt solution, and hatched as needed.

Provision of Foodstuffs: It's not enough to simply know and have on hand nutritious foods for cultured ornamental fishes. Their actual delivery in palatable formats, in sufficient concentration, at appropriate times is also critical. For the first few weeks of Clownfishes lives they require frequent (a dozen or more times per day) administration of nutritious foods where and when they metamorphose into small fishes, ultimately "settling", otherwise being more mobile than prevailing water movement.

Foods, either live and/or prepared should be fed several times daily, ideally on a continuous "as-needed" basis, and reciprocally never to excess. Various devices have been devised for automatically supplying these foods, from computerized peristaltic pumps to automated flake food feeders, to simply pinching low-fat dry foods between culturists' fingers. A survey of the literature finds that feeding per species, age and facility as few as three times to as many as fifteen times daily to cultured young.

Larval Rearing

Larval fishes have little tolerance for environmental stress; therefore culture systems must be made to provide optimized and stable conditions.

Metabolites, particularly ammonia have been shown to decrease vitality, resulting in slower growth rates and death. Water quality, particularly temperature and salinity are necessary to match with new water. This can be achieved through the use of large amounts of recirculated water or close matching through pre-mixed and stored change water.

As gametes, embryos and larval fishes suffer high mortalities in being moved, parent stock is often stripped or otherwise removed rather than their young. For brooding species like Clownfishes and Gobiids this point of moving the parents occurs when the young are free-swimming.

A Rearing Protocol for Gobiosoma Gobies is Given below as an Example: In anticipation of hatching, 10-gallon larval rearing tanks are filled with synthetic seawater (Instant Ocean) at a salinity of 30 parts per thousand (ppt), gently aerated through via an airstone. Subsequently 1 litre of concentrated *Isochrysis* along with approximately 40,000 rotifers, (enough for a concentration of 10 rotifers/ ml. are added. Post hatching, rotifer concentrations are estimated daily by removing 1 ml of water and counting individuals in a depression slide, using the 40x magnification dissecting microscope. 1 litre of *Isochrysis* is added to the tank daily, and rotifer concentrations are maintained at approximately 10/ml.

On or around day15 post-hatch, newly-hatched nauplii of *Artemia* are introduced to the diet. After a 5-10 day overlap period, the fish larvae are fed *Artemia* exclusively. All rotifers and *Artemia* are soaked in a commercially-prepared suspension of highly unsaturated fatty acids (HUFAs) for 12-16 hours prior to feeding.

Once the rotifer diet has been completely replaced by *Artemia*, *Isochrysis* alga is no longer added, and a pre-cycled air-driven foam filter is placed in the tank to help maintain water quality. Additionally, a 50% water exchange is performed every 3-4 days by siphoning water out through a section of flexible air tubing inserted into a 500-micron Nitex screen sleeve submerged in the tank. The sleeve prevents larvae from being siphoned out. Replacement water is then siphoned back slowly into the tank.

Gobiosoma reach metamorphosis around 30 days post-hatch. Metamorphosis is defined in these species by a rapid accumulation of colour (not useful as planktonic larvae) and settlement from a pelagic to a benthic mode of existence. Around the time of metamorphosis, dry feeds are introduced to the diet and becoming exclusive foods within 3 weeks. Juveniles are accumulated from various rearing tanks and consolidated into 29-gallon tank systems.

Grow Out Tanks: Though food-fish are cultured in open to semi-open systems, pumping fresh seawater to them and either allowing used water to flow back or to be partly (semi-open) filtered, aerated and recirculated for re-use, almost all ornamental marine fish culture involves closed or completely re-circulated water systems. The upside to closed systems is the greater control one has over initial and ongoing water quality. With such re-used water there is little chance of introduction of parasites, pests or pollution. The downside is the expense of its making and/or transport and manipulation (heating, filtering, pumping).

Water in grow out systems must be moved very gently. Particularly the first week or two of life is a delicate time with all but the finest of air-bubbles able to outright kill small larval fishes. Hence, either slow moving water introduced through fine screens and exited through fine screens (to keep young fishes in) or very fine air-stones are employed for water movement and aeration.

Some investigators have used as little as ten percent daily trade-out of culture water in larval grow out (Siddall 1979), but if good quality water of equal temperature and specific gravity can be assured there are many benefits that accrue from larger daily change-outs. Notable improvements from dilution of metabolites include improved growth rates, vitality and reduced intraspecific aggression.

Size of grow-out vessels is critically important. Due to problems of moving young fishes, these grow-out containers must be large enough to accommodate the batch of young through at least the first six weeks·or so of development. Houde (1973) proposes minimum volumes for some pelagic species being twenty gallons. Rationale for size is offered for allowing adequate gas exchange (at 3-7 larvae per litre, 11-27 per gallon), reducing aggressive interactions, diluting growth-limiting pheromonal metabolites (Yu 1968). For fishes of about Â½" total length a maximum density of two per gallon is suggested by Young (1995).

There are many designs of culture and grow-out containers, some involving elaborate curved tanks and screens. For many operations, simple rectangular glass aquariums of twenty or more gallons work fine, being readily available, inexpensive, easy to move and clean. Fitting these with a simple airstone and some mechanism for water changing works for most species under culture.

Maintenance: As per hobbyist use of synthetic saltmixes, these must be made up, aerated and stored for several days ahead of use to ensure overall stability. For closed-system arrangements, the water changed out from larval culturing systems can be used further for adult systems and food-culture facilities.

Often it is of benefit to keep phytoplankton blooms in the rearing tanks, such algae presence assuring improved survival and growth of larval fishes. The algae serve to remove metabolites, produce oxygen, diminish intraspecific interactions (possibly by simply decreasing visibility), feed zooplankton foods. Various genera have been employed for this purpose (*Chlorella* sp., *Chlamydomonas* sp. *Anacystis* sp.)(Siddall 1979). In actual practice, particularly when dealing with semi- to fully open systems, no specific measured amounts of concentrated algal culture are administered, but some added on a daily basis to render the water a light green in colour.

Larval fishes must be examined often and closely for signs of intraspecific aggression or its manifestations (dissimilar growth et al.) and increased water changes, addition of dÃ©cor (bits of coral, plastic to physically break up the environment) placed to reduce same.

The importance of avoidance of toxins cannot be overstated. Tanks should be cleaned and triply rinsed to assure removal of any contaminants. Workers should abstain from tobacco use, and be instructed on thoroughly rinsing of their hands and arms in advance of hatchery work.

To keep fishes off the bottom of culture vessels, feeding and therefore growing, some ambient lighting is supplied continuously in culture facilities. This is especially important in separating the young from the inevitable accumulated detritus on tank bottoms, and the ill-effects of high bacteria counts associated with it (loss of growth, health, life)(Young 1995).

Diseases, Pests and Abnormalities: Compared to the wild, captive-produced ornamental fishes have spectacular hatch out and larval survival rates. Nonetheless, infectious disease and genetic and developmental defects can play significant roles in incidental mortalities and loss of salability of stocks.

Certain bacteria (*Pseudomonas*, *Vibrio* spp.) are likely omnipresent in culture systems. Early efforts in culture of ornamental fishes employed antibiotics (e.g. penicillin, streptomycin, Chloramphenicol, sulfamerazine) to retard their growth. Later methods of bacterial population limitation include ultraviolet sterilization and ozone use.

Special care must be exercised in employing antibiotics in systems relying on biological filtration in closed settings, as these compounds may cause a loss of nitrification partially or wholly.

There is a variable potential of introduction of parasitic disease and pests with the use of wild-collected planktonic foods. So much so that almost all facilities engage their own food-production facilities, keeping close controls on contamination of stocks.

Early (1960's-80's) culture of marine ornamental fishes showed a definite resistance in acceptance by the trade all the way to consumers on the basis of wild-fishes larger sizes, greater colouration, and a decided percentage of captive fishes with physical deformities (bent mouths, shortened unpaired fins) and markings (incomplete bars on Clownfishes) due to environmental trauma (thermal and osmotic shock, high metabolites, contact with culture vessel walls). Happily, these abnormalities have been largely solved, and the demonstrated greater survivability of captive produced fishes has shown them to be much better purchases than wild-collected specimens.

Benefits of Captive-Produced vs. Wild-Collected Specimens: An assured shortcoming element of aquacultured fishes has been their apparent higher cost through the chain of custody/supply. What good are a ready supply of captive-produced species when wild-collected ones that appear the same or better colour, size, behaviour-wise are available at lower cost? With improvements in culture, changes in law (national and international), freight costs, and above all, sensitivity and awareness of retailers and hobbyists, there has been a groundswell movement to captive produced livestock. "Net-landed cost" of specimens, especially in consideration of "longer term" (months instead of days, weeks).

Similarly there are benefits in reduction of pressure on wild stocks extraction, avoiding corollary environmental damage during collection, loss of life of "by-catch", and very importantly, near absence of infectious and parasitic disease. Before the advent of cultured Clownfishes it is likely that some 95-58 % mortalities of wild-collected stocks occurred within a month of their capture.

Cloze: Many other species of fishes have been successfully produced in captivity, croakers (family Sciaenidae), most of them for human protein markets.

With increased interest in captive-produced ornamentals, progress in their food culture, gained knowledge in natural history and husbandry techniques there will likely be few "hold-outs" in the spectrum of successful captive-bred and reared fishes in the future. This progress will further assure captive success in succeeding generations by aquaculturists and enhanced longevity in aquarium keeping of these species.

It is hoped that with much more captive-adaptable (more readily feeding on prepared foods, absence of pathogenic disease, reduced shipping trauma, eliminated wild-collection damage) that captive-produced ornamentals will quickly grow to supplant wild-collected specimens.

Fish Disease

Argulus

Argulus or fish lice as they are commonly called are a very major threat to your fish's health. They can cause significant morbidity and mortality when heavily infesting your fish and they have been known to be the vehicle for other fish diseases. Argulus are a crustacean parasite in the subphylum Crustacea, which groups them together with prawns, shrimp, water fleas and others. The animals that are in the Crustacea group have semi-rigid to rigid chitin exoskeleton. The chitin exoskeleton will molt, or shed, as they get larger. Argulus are in the Branchiura class, which is group of Crustaceans that all have similar features. All of the Branchiura are fish parasites.

Argulus have a direct life cycle using only the fish as hosts. They prey upon freshwater fish and marine fish. Argulus can spend a large amount of time swimming around and mating occurs when the male and female Argulus are swimming. The eggs clusters are dropped on any convenient submerged item. After hatching the Argulus makes several metamorphic changes as it goes towards adulthood. The whole cycle takes between 30-100 days depending on the temperature. After hatching they must find a host within a around 4 days or they will not survive.

Argulus are one of the biggest parasites and are seen with the naked eye. They range in size from 5 to 10 mm in length. While they are easy to spot if you know what you are looking for, they are quite easy to miss when doing a skin scrape. You may see small dark spots on your fish and not realise that they are Argulus until they move. You can usually find them located behind the fins or around the head, in sheltered locations. You can spot them easier on the fins than on other parts of the body because they show up better against the transparent background.

The fish lice are oval in shape and flat. They can move quickly and you can sometimes spot them in the tank when they move from one host to another. If you try to net your fish, they will abandon the host and move into the free water.

Affected fish have patches of hemorrhagic and edematous affected skin, gills or fins. The parasite causes these injuries by attaching to the fish with its curved hooks and sucker. Its feeding apparatus further injures the host fish when it inserts the stylet into the epidermis and underlying host tissue causing hemorrhage. Argulus feed on the host's blood and body fluids. The feeding apparatus also releases

Common Fish Diseases

Columnaris: Grayish white marks or patches on the body of the fish or around its mouth are the first obvious signs of Columnaris. These patches appear like threads, especially in the vicinity of the mouth. Hence, this is often confused with another disease called "Mouth Fungus". The fins of the fish may show signs of deterioration, the gills get affected and slowly sores start appearing on the body. During a later stage the fish can hold its fins very close to the body and not spreading them. Columnaris disease is caused by bacteria, and is mostly a direct result of poor water quality. Sometimes, the shock of being introduced to the new aquarium leads to Columnaris disease in newly imported fish. Anti-bacterial medications will be enough if the disease is discovered and treated at its early stages. The disease will however invade the internal organs during the later stages and antibiotics are then required to cure it.

Gill Disease

Gill disease make the fish appear to be swimming in a great hurry, but it does not move at all. The fins flap very fast, and the fish gasp for breath at the surface, or lie listless at the bottom. The gills become swollen and discoloured. Bacteria, fungi and parasites in water can cause gill disease. The first step in the treatment is therefore an immediate improvement in water quality. Frequent water changes, conditioning and dechlorinating the water to ensure its quality are steps to take to threat and prevent gill disease. The chlorine in water often irritates the gills of the fish. Adding an anti-bacterial solution to the water may be needed to cure the fish.

Ick

Ick is a disease that causes your fish to get small, grain-like spots on their body. Ick is also known as White Spot Disease or Ich. The fish may try to rub itself against the décor or the hard surfaces in the aquarium, giving an impression of scratching an itch. In freshwater, this disease is caused by protozoa named Ichthyophthirius Multifilis and is very contagious.

The parasite spends a part of its life within the skin of the fish, where they form the white spots. These are the cysts. Once the cysts mature they fall down to the substrate and give rise to thousands of new free-swimming parasites that are free to prey upon other fish. An outbreak of Ick is very difficult to control if not noticed in the earlier stages. Since the parasite is released into the aquarium, the entire aquarium has to be treated.

One of the effective ways to treat it is by adding salt to the water. The salt should be added gradually, slowly building upto5-8 tsp per gallon / 4L. Different fish have different tolerance level for salt. In the case of more sensitive fish, it is better to keep the level at 5 tsp. per gallon / 4L. Using medications based on Malachite green and formalin is also a way to control this parasite, but Malachite Green may be harmful to scaleless fish.

Killing the parasite can be done only when the cysts have fallen down and the parasites are released into the water. Increasing the water temperature speeds up the life cycle of the protozoa. Keep in mind that increased water temperature may deplete the oxygen in the aquarium.

Dropsy

The common physical characteristic of this disease is a hollow or swollen abdomen. The fish's scales may stand out in these swollen areas. It is easy to mistake such fish for pregnant females. The fish appear listless and lose appetite. Viruses as well as bacteria can cause dropsy. Bad water or infected food is things that can lead to dropsy out brakes in your aquarium. Once the damage reaches the kidney, the condition becomes almost incurable. The disease must therefore be caught early if you should be able to save your fish. This disease is unfortunately very difficult to cure even if discovered early. Dropsy is however is not highly infectious. Dead fish should still be immediately removed from the aquarium; otherwise healthy fish may come and eat the dead one. Commercially available remedies do best.

Fin-rot

As the name suggests, fin rot causes the fins to turn opaque and they may even become blood-streaked. The erosion of the fin may continue till it reaches the base of the fin. At this point, the fish dies. The bacteria that cause this generally do not affect healthy fish. When fish are stressed, or when the fins have already been damaged due to excessive fin nipping, then the disease takes root. Commercially available anti-bacterial medicines have to be used to treat the fish.

Fungal Infections

The spores of fungus are always present within the aquarium, and can become infectious when the fish already have damaged gills or skin. Fungal infections are typically secondary infections. White, cottony growth on the body of the fish is the main physical symptom. Anti-fungal medicines are usually the best option to fight the infection. Most of the anti-fungal medications available on the market also fight bacteria, thus providing double protection.

Hole in the Head

As the name suggests, the fish develop small pits mainly in the region of the head. The holes are usually white in colour and have a yellowish mucous trailing from them. The fish become listless, lose colour and stop feeding. Hole in the head disease is cause by protozoa and is usually a secondary infection. Low levels of the protozoa that causes hole in the head disease live within the fish, and poor water quality and/or poor diet leads to the infection. The fish should be isolated in a quarantine aquarium and then treated. Direct injections or medicated food are some ways to treat the fish.

Pop-Eye

The name suggests the physical symptom. The fish's eye starts to enlarge and looks almost as if it is ready to pop out of the body. The cause could be bacteria. Too much dissolved oxygen in the water may also lead to this condition. There is no universal cure. Isolate the fish, and treat it with general anti-bacterial medicine.

Cloudy Eye

The eye becomes cloudy almost to the point of whiteness and the fish lose vision. The causes behind this disease can vary. An increase in parasites in the aquarium is the most common cause but severe stress, old age, or malnutrition can all lead to this condition. Treating this condition requires an investigation of water quality. Once the water quality is high enough, the fish will usually recover by them selves within 1-2 weeks. Thus, it is advisable to wait for 1-2 weeks before administering antibiotics.

Swim Bladder Disease

This mostly affects fish that have a globoid body shape. A loss of buoyancy is one of the first indications. The fish stays at the bottom of the aquarium, unable to rise, or float at the surface only. This is due to the fact that the swim bladder is what helps the fish to remain buoyant. Any damage to this membrane causes this disease. There is considerable debate about the cause of this disease and the best way to cure it.

Fish Lice

Small, white lice are apparent on the body of the fish and the fish rub themselves against the hard surfaces in the aquarium. Red spots or sores develop. The best way to treat fish lice is to remove the lice using a set of tweezers. Siphon the substrate daily to get rid of eggs. If the infestation is heavy, medication will be needed.

Nematode Worms

The anus of the fish may protrude a little, and a thin threadlike worm will drift out of the anus. The fish becomes listless and bloated. It refuses to feed. Anti-worm medication developed for aquariums are usually affective against nematode worms. Nematodes are commonly present even in healthy fish, but can turn into a problem when the fish becomes weakened due to some other reason.

Water Quality Induced Diseases

The most common among these are caused by an imbalance in the water chemistry. Ammonia poisoning is caused when too much organic waste is present in the water. If the pH of the water goes above 7.0 the ammonia turns deadly and toxic for the fish. The first indication is that the gills turn a different colour and the fish gasp for breath and become listless. The source of the waste has to be removed and the pH of the water should be reduced to treat this condition. Carbon Dioxide poisoning can be caused by too much of Carbon Dioxide. Good aeration and frequent water changes are the best way to treat and prevent this problem.

Nitrite or Nitrate poisoning shows itself by making the fish appear too bright. The fish remain panting for breath and hang just below the surface. Poisoning may also occur by using untreated tap water.

Constipation in Fish

Constipation is not that uncommon in fish and can, if left untreated, cause the death of the constipated fish. Some fish are more prone to constipation than others and it is hard to give any general rules about which fish that are inclined to get constipated and which aren't. It can however be said that fish with high narrow bodies such as angelfish and silver dollars usually are more susceptible to constipation then fish with other shapes.

The symptom of constipation is swelling of the body and abdomen and loss of appetite. Constipation is usually not caused by any diseases but rather by the diet the fish are fed. If the wrong diet is given to a fish for a prolonged time this can cause constipation and constipation can therefore often be rectified by improving the diet of your fish. Exactly what this means depend on the species of fish you keep in your aquarium. Predatory fish are best kept on one type of diet and herbivores on another, and so on. If a predatory fish is kept on an herbivore diet or vice versa there will be problems. One possible problem is constipation; others include various deficiency caused diseases. Search this site, the rest of the Internet and books to find information about the right diet for your fish and make sure to feed your fish such a diet in the future. It is also important to remember the importance of a varied diet. Not varying the diet enough is a common cause for constipation as well as for different deficiency related health issues. If the fish is given a varied diet suitable for the fish in question constipation should never occur.

If the constipation doesn't resolve itself after an improvement of the diet of the fish you will have to try other methods. Feed your fish flake or other dried food that has been soaked in glycerol, castor oil or medicinal paraffin oil to help remove the colon blockage that causes the constipation in your fish.

Costia

Costia is one of the many fish diseases that are caused by protozoa. It is a somewhat rare disease that is relatively easy to cure if your fish becomes infected by it. There are several treatments for this disease but it is of course preferable to prevent the protozoa from ever infecting you fish by keeping you aquarium clean and your fish healthy and well fed, while at the same time doing what is possible to avoid introducing diseases and parasites into the aquarium.

Costia has more or less only one symptom, it causes the skin of the infected fish to become cloudy and milky. There are as earlier mention several treatments, two that are very effective and one that often work. The two that are very effective do however have some potentially serious side effects. The less effective one without serious side effects is to raise the temperature in the infected aquarium to 27-28 °C / 80 - 83° F for a few days which usually cures the disease. This method is not recommended for serious Costia outbreaks as it is not 100% sure to cure to disease and it is preferable to make sure to cure the disease as fast as possible in severe cases.

The two sure ways to treat costia is with Copper (add 2mg per litre water) or Acriflavine (trypaflavine) (add 1ml per litre water). Acriflavine (trypaflavine) can cause sterility in fish and it is therefore very important not to overdose it. Copper on the other hand can poison your fish and potentially kill them if overdosed. It is worth mentioning that some fish are very sensitive to copper and this treatment method should never be used in aquariums containing such species. Read up on your species or ask in the LFS to find out if your fish are sensitive to copper.

Danger Signs with Your Fishes

Getting to know your fish and their natural behaviours well is always a good idea, since it will make it much easier for you to detect problems at an early stage. Treating health problems before they get a chance to bloom is much easier than trying to save severely affected fish. In this article, we will discuss a few common symptoms that aquarists should look out for in the aquarium.

In some cases, a "symptom" can be a perfectly normal part of your fish's behaviour and observing your fish while they are still happy and healthy is the only way of distinguishing alarming symptoms from normal changes. Reading books and articles about your species is also a good idea since it will give you a heads up on what to expect in the future. How does your particular species act when it is getting ready to spawn? How does it alter its colours to show submission? What can refusal to eat mean in many African cichlids?

Abnormal Swimming

Abnormal swimming is usually one of the first symptoms of something being wrong in your aquarium. The fish can for instance seem "drunk", swim in circles or simply drift with the current. It can sink to the bottom and be unable to raise itself, or float at the surface, incapable of swimming any further down. Severe health problems can even make a fish swim upside down with its belly pointing towards the surface.

Respiration and Gills

Gasping and laboured breathing is an important warning sign that should not be taken lightly. It can be caused by a health problem or by improper aquarium maintenance. Generally speaking, it should prompt you to increase aeration and carry out a series of frequent water changes. Hopefully, the problem will go away. If not, the fish is probably diseased. Different fishes have different respiratory rates and it is therefore impossible to say how fast a "normal" fish breathes. Observe your fishes carefully while they are still healthy and remember how they normally breathe. It should be noted that many fish species are capable of breathing oxygen directly from the air and it is perfectly normal for such species to swim up to the surface once in a while, gasping for air. Gill sheets that loose their normal colour and turns pale is another serious warning sign that should never be neglected. Most healthy fishes have bright pink gill sheets, but there are a few exceptions; the gill sheets of the popular Betta splendens are for instance pigmented.

Anorexia

Anorexia can come in many different forms and knowing the normal feeding habits of you fish are therefore imperative. Some fish refrain from eating completely when sick, while others eat much less than normal, or only certain types of food. It is also quite common for sick fish to take a piece of food into its mouth and then promptly spit it out. Anorexia can be brought on by a myriad of reasons; one of the most common ones being constipation. Constipation is especially common in fish with laterally compressed bodies where there is little room for the intestines. As the constipation proceeds, it will normally cause a significant swelling of the abdomen and can prove fatal.

Anorexia does not have to be brought on by disease; it can just as well be a sign of improper living conditions or an unsuitable diet. In some cases, simply increasing the water temperature will make the problem go away. In other cases, fish anorexia is completely normal and will go away of it self. Mouthbrooding species can for instance refrain from eating while brooding and female fishes can stop eating as they get ready to spawn since the large collection of ripe eggs inside them blocks their intestines.

Body Shape and Contours

Watching the body shape and contours will give you a lot of information about your fish. It is a good habit to set aside a few minutes every day to check your fish for changes in body shape, swellings, torn or clamped fins, and so on. Swelling of the abdomen can be a good thing when it is observed in egg-bearing or pregnant females, but it can also be a sign of constipation, internal parasites or bacterial infections.

Colouration

A fish that does not feel good will often dampen its colours. Loss of colouration can be brought on by virtually anything, from skin problems and internal disease to low levels of oxygen. Frightened or otherwise stressed fish will also normally look less vibrant than normally. Keep the aquarium well decorated, provide your fish with a wealth of hiding spots and keep your eyes out for bullies and fin nippers. Dampening the light can also prove beneficial.

A submissive male can voluntarily choose to dampen his colour and clamp his fins to avoid picking fights with a dominant male in the aquarium. In this scenario, the loss of colouration is actually a good thing since it will prevent violence and injury. Loss of colouration can also be associated with spawning; the fish can for instance be exhausted after raising a batch of fry or putting up with a lot of aggression during the pre-spawning period.

Cysts, Sores, Blood-Stains Etcetera

Diseases of the skin will often manifest in the form of cysts, spots, sores, slime, ragged white to brown patches, or small blood stained patches (so called echymoses). Examples of commonly occurring skin problems in aquariums are Ich, Velvet and fungus attacks. A fish with skin problems will often be itchy and scratch itself against rough surfaces in the aquarium.

Dropsy

Dropsy is a somewhat mysterious disease that is fatal and that can be hard to cure. The disease is caused by bacteria and (almost?) only attacks weakened fish. There are many reasons that can cause a fish to become weakened and most are within your control as an aquarist. The reasons include poor living conditions, incorrect living conditions, improper diet and incorrect company in the tank. By keeping your aquarium clean and well kept you can eliminate the risk of dropsy due to poor living conditions.

By researching the requirements of your fish and setting up the aquarium accordingly you can eliminate the risk for dropsy due to incorrect living conditions. By feeding a varied diet suitable for your fish you can eliminate the risk for dropsy due to poor nutrition and by only keeping fish with similar needs and temperament together you can eliminate the risk of dropsy due to incorrect company. This does of course not eradicate the risk of dropsy completely, but it does significantly reduce the risk of dropsy.

Dropsy is as earlier mentioned a bacterial disease, it attacks the kidneys on weakened fish and cause renal failure or/and fluid accumulation. It is the fluid accumulation that causes the visible symptoms of this disease; bloated body and protruding scales.

It can as I said be hard to treat bloat and many infected fish dies which makes dropsy a dreaded disease. There are medicines against dropsy that you add to the water of your tank but those are seldom very effective and it is, as this is an internal disease, better to give the fish medicated food. Make a 1% mixture of an antibiotic such as chloromycetin (chloramphenicol) or tetracycline and fish food. Based on weight, 1% equals to add 100 mg antibiotics to 10g fish food, or 25g fish food with a standard 250mg tablet antibiotics. It can be good starve the fish before feeding them medicated food to make sure that they eat properly.

Ergasilus

Ergasilus are parasites that infect fish and are very similar to the anchor worms both in regards to symptoms and treatment. These parasites can often weaken fish and thereby invite other diseases to infect the already sick fish.

Ergasilus can cause many of the same symptoms as anchor worms do, but are smaller and are more particular about the areas of fish they infect. The Ergasilus parasites infect the gills of the fish and can therefore besides the regular symptoms (such as the fish scratching itself against everything in the tank and green threads hanging out of the fish with an inflamed centre) also cause the fish to be sluggish and having trouble breeding. Fish can seem otherwise healthy and just display the white-green threads and still be infected by Ergasilus.

This disease or rather parasite can just as the anchor worm be treated using a potassium permanganate. Treatment can be administrated to the entire community/holding tank, which is messy since potassium permanganate will colour the water. Another less messy alternative is to bathe infected fish in a strong potassium permanganate solution. I recommend treating the entire tank even if it is messy as this guarantees that you treat all sick fish. Ergasilus has a long development time and seemingly healthy fish can therefore be infected as well. Use 2 mg potassium permanganate per litre aquarium water if you want to treat the entire tank. A stronger solution is required to kill the Ergasilus using a single 20-30 minute long bath and it is recommended to add 10mg potassium permanganate per litre water if you choose to use this method. Keep an eye up for follow up diseases if your fish have suffered from Ergasilus to make sure you can catch and treat them fast.

Chapter 4

Eye Problems in Fish

Eye problems in fish can have a long row of different causes from bacteria to poor nutrition. The symptoms of eye problems vary as much as the causes of the eye problems and include everything from cloudy eyes (which often are a result of bacterial infection) to exophthalmia aka pop eye and opaque lenses. Physical injury is also quite common. If your fish in some way have physically harmed its eye you can help it recover by moving it to a recuperation tank with antibiotics in the water to prevent infections. You should also try to identify the cause of the damage and see if you can prevent it from happening again.

Cloudy eyes are as mentioned above most often a result of a bacterial infection even if it in rare cases can be caused by fungus. Adding tetracycline, chloromycetin (chloramphenicol) or another antibiotic to the water or even better to the food can often (not always) help cure this problem. Start treatment as soon as possible to prevent the infection from causing serious injure or even blindness.

If your fish appears blind without any previous trauma or infection to it it's likely caused by excessive lighting or poor nutrition. Dimming the lighting and improving the diet can in these cases help prevent more fish from going blind but will not help the fish that has already gone blind.

Pop eye, formally known as exophthalmia, can be caused by vitamin A deficiency, tumours, bacterial infection, gas embolism and rough handling. Pop eye caused by vitamin A deficiency can often be cured by an improved diet with more vitamin A. Pop eye caused by tumours or rough handling can be hard to do treat and it is unfortunately often up to the fish itself to try to recuperate from. Gas embolism and bacterial pop eye can often be treated with penicillin or amoxicillin.

Opaque eyes are caused by metacercaria or nutritional deficiencies. Improving the diet and including a lot of vitamins in it can often help cure opaque eyes.

Fish Necropsy

Performing an autopsy (necropsy) on dead fish might sound grouse, but it is actually a very good idea. By examining your fish after its demise, you may be able to figure out what killed it, and if other fish develop the same symptoms in you aquarium, it will be easier for you to diagnose the problem and carry out the right treatment. If there is no treatment and the chances of spontaneous recovery are slim, you may wish to euthanize you fish to save if from further suffering and prevent the disease from spreading.

Even when you know what killed your fish it can be a good idea to examine it closely once it has died, since this will provide you with a deeper understanding on which conditions that are normally occurring in your fish and wish conditions that are not.

This is only a very brief article about fish autopsy. If you want to start examining your fishes, there are several detailed books and articles available that will provide you with more comprehensive guidelines. It is also a good idea to consult a comprehensive fish disease book since a myriad of different organisms can infest a fish and they can be tricky to tell apart for the untrained eye.

If a threatened or endangered fish dies in your aquarium, you should ideally contact a qualified veterinarian or fish export to perform the autopsy for you. If there is a breeder organisation or similar, you can contact them for further advise.

External Autopsy

A fish autopsy will normally start externally. Use a binocular to carefully dissect eyes, skin and fins. Keep an eye out for external lesions and bleeding. If you have access to a microscope, you can use it to check external scrapings. A rich profusion of different organisms can cause external damage to fish, including parasites such as Argulus, isopods, copepods, monogenetic trematodes, parasitic dinoflagellates, and various protozoa.

When you have checked eyes, skin and fins, it is time to cut off the tail and drain the blood from the gills. The goal is to remove the gill arches without leaking any blood that might obscure the gills during the inspection. The autopsy should ideally be carried out as soon as possible when you notice a dead fish in the aquarium, because if you do it early, most gill parasites will still be alive and easy to detect since they will continue to move around.

Before you start cutting the fish up for the internal part of the autopsy, carefully check the mouth and throat of the fish and look for abnormalities.

Internal Autopsy

When you proceed to examine the inside of your fish you will often stumble over organisms that may seem dangerous, but that had little to do with the death of your fish. It is true that internal disease can cause fish death, but it is also true with that a lot of organisms that live inside fish, such as internal worms and protozoans, the host and the parasite are to well-adapted to each other that they can co-exits until the fish dies of other reasons.

The first step of the internal examination is normally to carefully cut the fish from the vent to the isthmus to get to the internal organs. Be very gentle when you cut and make sure that you do not damage any internal organs or major vessels. Inspect the fish for internal bleedings, such as free blood in the coelom, damaged vessels, or bleeding organs. Also keep an eye out for tumours, blood clots, blocked ducts and vessels, a ruptured or punctured digestive tract, and lesions. Any organ that looks strange, e.g. that is enlarged or discoloured, should be inspected in detail. As you can see, prior knowledge from "healthy" fish is important; otherwise it will be difficult to notice abnormalities. Internal abnormalities in aquarium fish are often the result of bacteria, tumors or a blocked duct. Internal worms that normally cause no problem can prove fatal if they migrate through a vital organ or get stuck in a duct or vessel.

Fish Tuberculosis

Causes and symptoms of fish TB. *Mycobacterium marinum*. Possible treatment. There is some danger to humans when servicing

Tuberculosis was once a dreaded disease in Europe as well as North America and virtually everyone knew someone who had succumbed to 'consumption', the commonly used name for the disease caused by *Mycobacterium tuberculosis*. Tuberculosis is still widespread in most third world countries and after the fall of the Soviet Union, the numbers of tuberculosis patients have sky rocketed in Russia as well as in many other post-soviet states.

What few of us know, however, is that a bacterium closely related to *Mycobacterium tuberculosis* called *Mycobacterium marinum* can be present in wild caught as well as captive bred fish and stay lurking in our aquariums without us ever realising it. After all, when one of our fishes goes belly up, few of us drag out the Petri dish and starts growing bacterial cultures to find out exactly what caused its demise.

Since *Mycobacterium marinum* is so closely related to *Mycobacterium tuberculosis*, the disease it causes is commonly referred to as fish tuberculosis of fish TB. Learning more about fish tuberculosis is recommended for all aquarists since it can be lethal to our beloved fish. What's even worse, *Mycobacterium marinum* can spread to the aquarist and cause serious health problems. It can also spread to other animals and is capable of surviving in both soil and water without any host for prolonged periods of time. Don't be fooled by the word "marinum" – this nasty little organism will survive just as well in freshwater conditions.

Susceptibility

Certain types of fish seem to be more at risk of carrying, or at least succumbing to, fish tuberculosis. This group includes the labyrinth fishes, among which you will find many popular aquarium fishes such as Bettas and Gouramis.

Symptoms in Fish

The main symptoms of fish tuberculosis are loss of scales, loss of colour, lesions on the body, wasting, and skeletal deformities such as curved spines.

Diagnosing

Looking a slides of infected tissue under a microscope is sometimes enough to recognise Mycobacterium marinum, but in most cases a bacterial culture will be necessary. Both Mycobacterium tuberculosis and Mycobacterium marinum are acid fast, which means that they stain bright pink against a blue background.

Treatment

Treating fish tuberculosis is really difficult and euthanizing the fish a probably less painful for the fish than forcing it to go through endless treatments that may not have any effect on the disease. Euthanizing all the fish in the infected aquarium is also the best way of preventing the disease from spreading.

If you decide to try and treat your fish, keep in mind that Mycobacterium marinum can infect you as well. The risk of being infected can however be decreased dramatically by following a few simple safety guidelines. You can read more about this further down in this article.

Fish can be treated with the same drugs as humans get when they become infected by Mycobacterium marinum, e.g. Kanamycin. Since this is a very resilient microbe, normal treatment involves administering at least two different medications over the course of at least three months.

A lot of aquarium problems can be fixed by performing frequent water changes, increasing the water temperature and adding some salt to the water, but fish tuberculosis is not one of them. Raising the water temperature may even worsen the problem since Mycobacterium marinum prefers warm water (their ideal temperature is 30°C).

Prevention

Since curing fish is virtually impossible once the disease begins to manifest, preventative measures are highly important.

Keeping your fish healthy, happy and well-fed will boost their immune system and make it possible for them to handle limited exposure to Mycobacterium marinum.

Wounded or otherwise weakened fish should be moved to quarantine tanks where they can be treated and given time to recuperate, since weak fish that is left in the main aquarium can serve as a breeding ground for all sorts of malicious microorganisms that may eventually grow numerous enough to attack even the healthy fishes.

New fish should ideally be quarantined before you allow it into you main aquarium. Plants, substrate, equipment etcetera should be sterilized to kill of potentially harmful bacteria before being introduced to the aquarium.

An aquarium that has had an outbreak of fish tuberculosis should be meticulously cleaned out with bleach and left to dry before you restock it.

Fish Tuberculosis and the Aquarist

Prevention

When aquarists become infected by Mycobacterium marinum, it is usually because the carry out maintenance work when they have cuts or other skin problems on their hands or arms. Our skin is remarkably good at keeping malevolent microorganisms out, but as soon as the skin gets injured, an important part of the body's defence system has been breached.

It doesn't have to bee a large wound; a simple paper cut or eczema can be enough for Mycobacterium marinum to slip through. When handling an aquarium where you suspect that Mycobacterium marinum may be present, it is consequently important to use protective gloves. You may have a tiny sore that you haven't even noticed, such as a torn cuticle. Some aquarists prefer to use gloves all the time, or at least when they have damaged skin, since it is impossible to know if Mycobacterium marinum exists in an aquarium before the fish start to show symptoms of fish tuberculosis. You can also catch fish tuberculosis by using your mouth to start a siphon. Washing your hands and lower arms with soap after handling fish and aquariums is naturally always recommended.

Symptoms and Treatment

When aquarists become infected with fish tuberculosis, it normally starts as a skin problem. Mycobacterium marinum is a slow growing mycobacterium and it can therefore take several weeks before you notice any symptoms. The first sign is normally small purple lesions or "bumps". The lesions will grow and spread and Mycobacterium marinum can proceed to destroy the soft tissue under the skin, including tendons and joints. In severe cases, fish tuberculosis can spread to the bones and cause symptoms similar to arthritis.

Most doctors never come in contact with fish tuberculosis and if you show up with some strange lesions on your hands, they might not realise what is causing it and put you on a general oral antibiotic that will not kill Mycobacterium marinum. It is therefore very important to inform your doctor that you are an aquarist and that you may have caught fish tuberculosis or some other disease from your aquarium. Mycobacterium marinum can be successfully treated but only if the right combination of drugs are used. Just as with

Mycobacterium tuberculosis, the treatment will normally need to be carried out over the course of several months. In serious cases, intravenous administration of antibiotics may be required.

Flukes

Flukes are not the name of a single disease but rather the name of a number of parasites. The term flukes are used for a number of different about 1 mm long flat worms that all can be treated and diagnosed in the same way. Unlike many other health problems in fish, flukes can infect healthy fish if introduced to your aquarium and it is therefore important to try and prevent them from ever reaching your aquarium in the fist place.

Preventing flukes from entering your aquarium can be done by quarantining all new fish for a couple of weeks before letting them into your main aquarium and by avoid using live food from insecure sources such as water with fish in them. Ideally you should avoid feeding live food that you haven't cultivated yourself to make sure that you avoid the introduction of flukes with the live fish food.

The symptoms of flukes are similar to those of ich/ick. It is however possible to see the difference using a handheld magnifying glass. If you are dealing with flukes you should be able to see eyes and movement which you won't see if you are dealing with ich/ick.

Common symptoms include fish scratching itself against everything all the time, mucus on gills or body, sunken in bellies, rapid breathing, and reddened skin. In severe cases the fins on infected fish can be eaten clean of the fish. Flukes will if left untreated kill the fish by destroying its gills.

Fluke treatment is carried out using potassium permanganate. You can either bathe the sick fish in a potassium permanganate solution (10mg per litre water) for about 20-30 minutes or treat the entire holding tank.

Treating the entire tank will help make sure that you kill all flukes and help you avoid a new fluke problem, but it will be a little messy as potassium permanganate will colour the water in the aquarium. Despite this I recommend the later option as it is safer and doesn't involving catching the sick fish and subjecting it to extra stress, and because it helps make sure that all flukes are eradicated.

Fungus in Fish (Saprolegnia)

Fungus is as the name tells us caused by fungi (Saprolegnia) and will only attack weakened fish that has already been infected by another disease. Fungus will never attack an otherwise strong and healthy fish and the best way of avoiding this disease it therefore to keep your aquarium clean and your fish healthy and well fed.

Fungus will eventually kill infected fish when it has spread through the fish even if this often takes quite a while to happen. The symptoms of fungus are gray or white growth on the fins and skin of the fish. The gray or white areas grow larger and more cotton like as the disease progresses.

Fungus is fortunately rather easy to treat and there are several different ways to treat it. One is to add methylene blue to the water. Methylene blue are most commonly used to prevent fungi on fish eggs but can be used on adult fish as well even if it isn't necessarily the best option for adult fish.

Personally I recommend using methylene blue for treating and prevent fungi on eggs and fry. On adult fish I instead recommend using a phenoxethol solution or commercial antifungal medicine designed to treat fungus in fish.

If you choose to use phenoxethol solution you should add 10ml of 1% phenoxethol solution per litre aquarium water. The treatment can be repeated after a few days if needed but should not be done more than twice since repeated use can have damaging side effects. Severely infected fish can be caught and swabbed with a povidone, iodine or mercurochrome soaked cloth to increase the effectiveness of the treatment. Make sure to increase the water quality and fish health in your aquarium once the fungus has been cured to avoid the problem in the future.

What is Ich?

Ich is caused by a ciliated protozoan of the genus Ichthyophthirius – Ich is simply an abbreviation of the word Ichthyophthirius. Some people prefer to spell it Ick, perhaps because they find the disease icky? Ich is also known as White Spot Disease since it causes small cysts to form on the skin of infected fish. If you look at the protozoan under a microscope, you will see a round rolling mass with a u-shaped nucleus. This parasite is completely surrounded by cilia.

The Lifecycle of Ichthyophthirius

The Ich parasite starts its life as a free swimming organism that moves around in the aquarium, looking for a suitable host. When it finds a fish, it will attach itself to the outer layer of the fish's skin and start feeing on its bodily fluids.

It can also attach itself to the gills, thus making it hard for the fish to breathe. Soon, the parasite will have developed a protective outer shell, a so called cyst, and it is this shell that we can see as a small white grain on the skin of infected fish. A fish heavily infested with Ich parasites can look as if someone had sprinkled it with grains of salt.

In the encysted stage, the parasite is a so called theront. The theront stage will end when the organism has grown big enough to break through its protective shell and drop to the bottom of the aquarium. It will attach itself to any surface, e.g. gravel, plants, tubing, or glass, and start to reproduce. The theront has now turned into a trophozoite (also known as trophont). Within 36 hours, one single trophozoite can produce hundreds of free-swimming Ich organisms – so called tomites – and the life cycle will start over again. Only the tomites are able to infect fish and only the tomites are sensitive to treatment.

Ich in the Wild

Ich is commonly occurring in nature but it poses less of a threat in its natural environment, due to two main reasons. Firstly and foremost, it is much harder for the parasites to find suitable hosts when they have to search a huge body of water. In a small aquarium, virtually every single parasite stands a high chance of stumbling upon a susceptible host.

Secondly, there are more fishes for the parasites to infest in the wild. The risk of one single fish becoming infested over and over again is therefore much lower in a large body of water than in a small aquarium. In a closed system like an aquarium, the parasites will infest the same poor fishes over and over again, gradually weakening them and thus making them even more susceptible to new Ich attacks.

Thirdly, a fish that manages to combat Ich seems to develop some degree of immunity to the parasite. Wild fish that becomes infested by just a few parasites have a greater chance of fighting them off, thereby developing a higher resistance towards new attacks. In an aquarium where a fish might be attacked by hundreds of parasites at once, its chances of successfully fending them off and developing immunity is much less

Preventing Ich

Completely eradicating Ich from your aquarium is not easy, and studies show that small amounts of Ich tend to be present even in well maintained aquariums where the fish are in prime condition. Just like the immune system of a healthy person can handle a few germs in the air, healthy fish seem to be able to live with Ich around them without falling ill. If we on the other hand crammed that healthy person into a subway and forced him to stand in a crowded wagon filled with sneezing persons, it might soon be too much for his immune system to handle and he would catch a cold.

The same thing is true for fishes – if you place them in a crowded environment, rarely perform any water changes and introduce a bunch of diseased specimens, you may very well have an Ich outbreak on your hands. The best way of preventing Ich is therefore to avoid crowding, perform sufficient water changes and quarantine new fish before you allow them into your main aquarium.

In addition to the causes described above, factors that are detrimental to the immune system of your fish can easily cause an Ich outbreak and must therefore be avoided. This includes keeping your fish in a barren aquarium where it feels stressed, keeping it with bullying tank mates, and keeping schooling fish alone. Water temperature and water chemistry should always be maintained within suitable parameters, and do not forget that many species handle rapid changes poorly. Even changes that are within the recommended water parameters can prove harmful to such species. Keeping the levels of organic waste down in the aquarium is also very important if you want your fish to stay healthy.

A lot of fish species can survive in an unsuitable environment, e.g. a barren aquarium where water changes are hardly ever carried out, but it will be unfavourable for their immune system and sooner or later they will probably succumb to a disease such as Ich.

Last but not least, fish that is weakened by other health problems are an easy target for parasites. It is therefore a good idea to move injured and diseased fish to a recuperation aquarium where they can heal. If they turn into a breeding ground for parasites and bacteria, at least they will not infect the rest of your fish.

Treating Ich

As long as the Ich parasite is protected inside its cyst, it is virtually impossible to get to - the levels of medication that would be necessary to affect the parasite inside the cyst would kill the fish as well. This is why we must focus on the free-swimming tomite stage when the parasite is much more vulnerable.

- Bring up the water temperature to 85-88 degrees F (if you think you fish can handle it).
- As mentioned above, Ich can attach itself to the gills of fish and make it hard for them to breathe. It is therefore a good idea to increase aeration in the aquarium to keep the levels of oxygen really high. Adding more aeration is especially important if you increase the water temperature since cool water holds more oxygen than warmer water.
- Add roughly one teaspoon of salt per gallon of water. (You can adjust this amount depending on how salt tolerant you fish are.)
- Carry out a series of water changes and clean all the gravel. Changing around 50% of the water once a day is recommended.
- Continue the treatment for at least one week, since you can only kill the free-swimming Ich parasites. The speed of the life cycle is temperature dependant (it will for instance take three days at 80 degrees F) and continuing treatment for an entire week is recommended to be on the safe side.
- If this is not enough to combat the Ich outbreak, you need to visit your local fish store and pick up some anti-Ich medication. Unfortunately, the Ich parasites seem to become more and more resilient towards treatment each year. Lazy aquarists and pet shop keepers are often tempted to constantly use medications to fend off disease, instead of devoting themselves to frequent water changes etcetera, and this makes it easy for resistant strains of parasites and bacteria to develop.

Glugea and Henneguya

Glugea and Henneguya are two different fish diseases that I due to their rareness and similarity will treat in one article. The diseases are as I say very rare and most aquarists will never encounter them but it can still be a good idea to know about them just in case as there are no known cures for these highly contagious and deadly diseases.

Glugea and Henneguya are like most other diseases most likely to infect weakened fish and keeping a clean aquarium with healthy and well fed fish will therefore help prevent Henneguya and Glugea. Always quarantining new fish for a couple of weeks will also help prevent the introductions of these diseases and so will only feeding live food from trusted sources or avoiding live food all together.

The symptoms of these diseases are very similar and it is hard to tell one from the other without access to a laboratory. Henneguya and Glugea both cause white cysts similar to those caused by Lymphocystis on the fins and body of the infected fish. In more sever cases the entire fish will bloat up.

As it can be hard to tell the diseases apart and the fact that Lymphocystis unlike Henneguya and Glugea isn't fatal, it can be a good idea to keep infected fish alive for a while to determine if the fish is suffering from the fatal Henneguya and Glugea or the less serious Lymphocystis virus. You should however make sure to quarantine suspected cases of Glugea and Henneguya immediately to prevent that the deadly disease spread to other fish. It can also be good to disinfect the aquarium once the diseased fish has been removed to further help reduce the risk of the disease spreading. Disinfecting can be done by adding disinfectants such as acriflavine (trypaflavine) or monacrin (monoaminoacridine) to the aquarium.

Hexamita

Hexamita is caused by intestinal flagellated protozoa and attacks the digestive system and the lower intestines of infected fish. The symptoms of this disease are often confused with those of hole in the head diseases as these two diseases usually are present at the same time in a fish. They are however two separate diseases that each needs to be treated. Some fish species are especially prune to Hexamita such as large cichlids and especially Oscar and discus fish.

The first sign of Hexamita is usually that the faeces become whitish and slimy. During this stage of the development of the disease the fish can still appear to be healthy and eat as usual. As the disease progress the numbers of symptoms will increase and you can often start to see symptoms of hole in the head disease as well. The symptoms of more progressed Hexamita includes the fish becoming darker or even black, the fish start swimming backwards, the head sinks in (becomes thin), weight loss and the fish starting to hide in small corners with its head facing downwards. Hexamita does as earlier mentioned attack the digestive system and can therefore cause nutrient deficiencies in the infected fish. Nutrient deficiency is one of the reasons behind hole in the head disease.

Since the disease is internal it is necessary to medicate internally. The easiest way to do this is to mix 1% metronidazole in fish food (25g food for one 250mg metronidazole) and feed it to the fish. Feeding sparsely before feeding the medicated food can help guarantee that the fish eats the medicated food. It is also recommended to medicate the water as well by adding 12 mg metronidazole per litre aquarium water to make the treatment more effective. Repeat the water treatment every other day until you have treated the water three times.

Hole-in-the-Head Disease

Also Known as: Head and Lateral Line Erosion Disease (HLLD or HLLE), Lateral Line Erosion (LLE), Lateral Line Disease (LLD).

This disease is especially common in discus fish and Oscar fish and causes large cavities and soars on the head and along the lateral line of the fish. The open wounds caused by hole in the head disease are very distinctive and makes this disease easy to discover and diagnose. If your fish has growing wounds on its head it is likely suffering from hole in the head disease. Early signs of this disease are small pits on the head of the fish. This disease often appears simultaneous with a disease called Hexamita and this has lead to these two diseases being frequently confused with each other. Hexamita attacks the intestines of the fish and can often be the cause of the hole in the head disease which has been attributed to nutritional deficiency, The two diseases are however not one and the same.

Hole in the head disease is believed to be the result of a deficiency of one or several of the following nutrients: Phosphorus, Calium, Vitamin C and Vitamin D. This deficiency can as earlier mentioned be caused by Hexamita and its effect on the digestive system but it can also be a result of poor water quality, poor diet or over filtration using chemical filtration.

Hole in the head disease is often very hard to cure, especially in its later stages, but an improved diet with more vitamins and increasing the water quality by cleaning the tank and performing more frequent water changes might help. It is also recommended to remove any chemical filtration such as active carbon filtration. Improving the diet of the fish can include adding vitamins to the diet, introducing vitamin rich flake food to the diet and removing (or at least cut down on) nutrient poor feeder fish and beef heart from the diet.

Ichthyosporidium

Ichthyosporidium is as the name suggests a fungus. It will as most other funguses only infect already weakened fish and will normally not be able to infect healthy fish in good condition. Making sure to keep good water quality, with the right water parameters for your fish and giving them a good varied diet will therefore prevent this disease from ever being a problem. Disease prevention is always preferable to disease treatment and this is true for Ichthyosporidium as well.

The Ichthyosporidium fungi attacks the fish internally and can because of this be harder to treat than many other fungus diseases. You can find more info about treating the disease later on in this article, but I want to start by looking at the symptoms of Ichthyosporidium. The disease does as earlier mention attack the fish internally and primarily attacks the liver and kidneys even though it will eventually engulf the entire body of the fish. This can cause a wide variety of symptoms in different fish but the symptoms usually includes the fish becoming slower, their bellies sinking in, the fish loosing balance and eventually the production of cysts and soars on the body. If the fish develops cysts or soars, it is very difficult to save it.

Ichthyosporidium is as earlier mentioned hard to treat as it is internal and the available treatments can cause damages to the fish and potentially even kill a fish that has already been weakened by Ichthyosporidium. These treatments include adding Chloromycetin or Phenoxethol to the food given the infected fish. Add about 1% of these substances to the food. 1% equals mixing one 250 mg pill with 25g food. The medicine can be mixed in most types of food. It is good to skip a feeding or two before starting to feed the medicated food to the fish to make sure that the sick fish eats the food. There is no guarantee that the treatment will work.

Diagnosed fish should be removed from the community aquarium as soon as possible to prevent the disease from spreading since it can be hard to cure. Some recommend destroying any fish diagnosed with this disease immediately but that can be a bit drastic as it might be possible to save your beloved fish. I do however recommend killing fish displaying cysts and soars as it is unlikely that those will recover regardless of treatment and you should always kill infected fish rather than letting the disease spread if you are unable to move sick fish to a treatment tank.

Injuries in Fish

There are a lot of reasons as to why fish might be physically injured in your aquarium and most such injuries heal by themselves without you having to worry about them. Lighter injuries are for example very common in aquarium with aggressive territorial fish where fights are a part of everyday life. More serious injuries can however need your attention to prevent the wounds from becoming infected and to prevent diseases from infecting the fish while it is weakened by the wound.

The first thing you should to when spotting a more serious injury to one of your fish is to isolate that fish in a quarantine/recuperation tank where it can heal from its wound without being stresses or further injured by other fish. It can also be good to pad the wound with an antibiotic such as Mercurochrome or to add an antibiotic like tetracycline and chloromycetin (chloramphenicol) to the water in the recuperation tank to prevent the wound from becoming infected. A lower pH will help speed up the recovery if your fish tolerate lower pH values. A pH around 6.6 is ideal for healing.

When you have tended to your fish you should try to identify the cause of the injury. The cause can be a sharp piece of decoration, another fish or a myriad of other causes. Some injuries are nothing more than accidents and have no real cause that can easily be prevented. When you have established the reason behind the injury, the next step is naturally to consider what can be done to avoid future damages. If it is another fish you might want to consider not keeping those two fish in the same aquarium, if it is a piece of decoration I recommend removing it or burying the sharp side in the gravel (unless you have digging fish).

Leeches in Aquariums

Leeches are parasite that are easy to spot and identify as they tend to be large enough to be spotted with the naked eye. Leeches are worms that often appear hearth shaped as they curle themselves toward the body of the fish. Leeches attached themselves onto the body, fins or gills. Leeches can only be found in your aquarium if you have introduced them to your aquarium somehow.

The main way to introduce leeches is with live food that has been caught in leech infested waters. Refraining from feeding live food you have not cultivated yourself and especial wild caught live food will dramatically reduce the risk of getting leeches in your aquarium. Leeches can also be introduced with new fish and plants. You should therefore carefully examine new fish and the water in the bag for leeches before letting the fish into you aquarium. Ideally you should also keep the new fish quarantined for a while to prevent introducing other diseases. Plants should also be inspected to avoid introducing leeches, and to be extra safe you can place new plants in a 5 mg/l potassium permanganate solution for about an hour before adding them to your aquarium. The potassium permanganate will kill any leeches that might hide among the plants in question.

It might be tempting to get a pair of pliers and manually remove the leeches from you fish but you should never do that. This is due to the fact that the leeches borrow themselves into the surface of the fish and forcibly removing them can cause serious damage to the fish. You should instead bath the fish for 12-15 minutes in a 2.5% salt solution. This will most likely cause the leech to release its grip on the fish and if it doesn't, it should at least have loosened it enough for you to be able to remove it using pliers without damaging the fish seriously. Never use this method on salt sensitive fish such as rays and certain catfish species. For salt sensitive species or if you prefer to medicate your entire tank (to make sure to remove all leeches) I recommend adding 0.25 mg Trichlorofon for every litre water in your aquarium If you choose to treat the entire tank I recommend digging up your plants and using the earlier mentioned method to kill any leeches that might be hiding.

Lymphocystis

Lymphocystis is a non lethal virus disease that can infect most aquarium fish. The disease is rare and most aquarist will never have any problems with it. The disease is caused by a virus and can only infect your fish if it is introduced to your aquarium. The risk of introducing this virus as well as other diseases can be reduced by keeping new fish in quarantine for a few weeks before adding them to the general population. It is also recommended to never add waters from other aquariums to your aquarium as this can introduce disease if you are unlucky. The Lymphocystis virus and other fish diseases can also be introduced with live food and avoiding feeding live food or only feeding home cultivated live food can therefore further reduce the risk of Lymphocystis and other diseases.

The Lymphocystis virus causes small white swellings on fins and bodies of infected fish. There is no cure for this disease but most fish recover from the disease by themselves. This makes Lymphocystis one of the least serious diseases that your fish can be infected by. Your main concern if you discover Lymphocystis in your aquarium should be containment as there is no cure. Move all sick fish to recuperation/ hospital tanks. Keep the fish there until they have recovered 100%. It is true that the disease usually isn't fatal and that the fish usually recover but this doesn't mean that you shouldn't try to contain the disease as the temporarily lowered health status of the infected fish can open the door for other more dangerous diseases.

Marine Velvet

Marine velvet is as the name suggests a disease that is found in saltwater aquariums and it is one of the most common illnesses in saltwater aquariums. It can infect fish in all different types of saltwater aquariums from species tanks to reef tanks, from coldwater tanks to tropical tanks. Marine velvet is found in all the oceans of the world and is common in wild-caught fish. The disease infects stressed or weakened fish and is therefore common in newly transported and recently caught wild fish. The disease only seldom affects healthy fish but can infect healthy fish if they are kept in an aquarium with an already infected fish.

Marine velvet progresses fast and can quickly kill fish; it is therefore very important to spot and treat the disease as fast as possible. The symptoms include quick breathing (the fish breathes faster than normal) and "dust" on the body of the fish. This dust can be a number of colours but is usually white, grey, brown or yellow. Observe that dust doesn't mean spots; white spots on the fish are instead a symptom of the other common marine disease - saltwater ich. Other symptoms of marine velvet are fish scratching themselves against the aquarium decoration and loss of apatite. The disease usually kills fish by damaging their respiratory system. Marine velvet is very contagious and is fatal unless treated.

There are several methods to threat this marine velvet. One is to move infected fish to a hospital aquarium where the salinity is lowered. This will cure the fish as marine velvet can't survive low salinity. It can however put a strain on the fish and it can be hard to reintroduce the fish to higher salinity levels. Some fish species are also less tolerant than others of low salinity. You have to be able to accurately measure and manipulate salinity to use this method. Another option is to use copper to treat the disease. If using a copper based medicine you should always follow the instructions and make sure that you don't keep any organisms that are sensitive to copper in your aquarium.

Nematodes

Nematodes are threadworms that can infect all types of fish. They have to be introduced to the aquarium somehow to infect your fish, and are usually introduced with new fish, live food or plants. Nematodes can be hard to treat and it is therefore very important to prevent ever introducing the worms into your aquariums in the first place. This is not hard to do and if you follow the advice below you will likely never see nematodes in your tanks.

To avoid introducing nematodes with new fish you'll need a quarantine tank where you can place your fish for three weeks to observe them before moving them to your main aquarium. If the new fish has nematodes you will hopefully spot it and avoid introducing it to your main aquarium/aquariums. Plants can be given a bath in a parachlorometaxylenol and water solution to kill any nematodes that might hide among them. Preventing nematodes to enter your aquarium via live food is harder and the only way to be 100% sure is to only feed live food that you have cultivated yourself. If you still want to feed your fish outside live food I recommend only getting live food from reputable sources and from waters without fish in them.

The threadworms infect the entire body of infected fish but the only visible sign of the infection will be worms hanging out of the anus of the infected fish. Severe cases can cause the belly of the fish to sink in but most cases don't cause any external signs other then the anal worms and usually don't cause any real problems for the fish. It should be treated none the less.

Nematodes are as earlier described hard to treat but there are a couple of methods that might or might not be successful. One is to treat the fish using parachlorometaxylenol.

The treatment can be given by mixing the parachlorometaxylenol in food, as a bath or in the aquarium in which case you should use 10 ml parachlorometaxylenol per litre aquarium water. Soaking the food in parachlorometaxylenol and feeding it to the fish is usually the most effective alternative if the fish will eat the soaked food. The second method to treat nematodes is to feed you fish medicated fish food containing thiabendazole. Such food can however be hard to find and hard to get the fish to eat.

Neon Tetra Disease

The neon tetra disease is a disease that infects neon tetras and for which there is no known cure. The disease is caused by a sporozoan named *Pleistophora hyphessobryconis* and is usually fatal. The disease is sometimes called "Pleistophora".

The neon tetra becomes infected when spores enter the fish and the two main symptoms are dull colouration and restlessness in the fish. Fish are often infected by eating on dead fish and it is therefore very important to remove dead and sick fish as soon as possible. As the disease progresses, cysts start to develop and this will often cause the body of the fish to become bumpy. At this stage the Neon tetras will usual have hard a time swimming. The spine of the fish often becomes curved in the final stages of the disease before death.

There is as earlier mentioned no cure for this disease and preventing the parasite from ever entering you aquarium is therefore very important. The most common ways of introducing this disease to your aquarium are with live food and new fish. Avoiding the use of live food or cultivating your own live food dramatically decreases the risk of introducing *Pleistophora hyphessobryconis* to your aquarium. Brine shrimp is for example an excellent live food that can be hatched at home. Put new fish into a quarantine aquarium before releasing them into the main aquarium to avoid introducing neon tetra disease this way. There are reports that using diatom filters decrease the risk of this disease but it is not enough to eliminate the risk of the disease on its own.

There is another disease that gives similar symptoms that is often called "false Neon Tetra disease" and that disease is caused by bacteria. It is all but impossible to tell the difference between the two diseases without access to a laboratory.

Preventing and Treating Fish Parasites

Knowledge is power, and this is true even in the field of preventing and treating fish parasites. As a fish keeper, it is always a good idea to familiarize yourself with the various common fish parasites that you might encounter and learn more about how to detect them in your aquarium. Parasite infestations that are observed at an early stage is usually much easier to cure than parasites infestations that the aquarist does not notice until all the fishes are severely infested. By learning more about fish parasites and how to recognise them you can also decrease the risk of bringing home infected fish from the fish shop.

A lot of factor will play a role when it comes to determining the best cause of action when one suspect parasites in the aquarium. The species of parasite is naturally of imperative importance, but the fish species in the aquarium must also be taken into consideration since some fish species are exceedingly sensitive to certain parasite medications.

You must also decide if you want to treat the entire aquarium, or if it is better to move the fish to a separate hospital aquarium. In some situations, dipping the fish in a bowl of medication will be sufficient. The volume of your aquarium is also a factor to take into account since some medications are very expensive. (Sometimes euthanizing and replacing all the fish will be cheaper than trying to cure them.)

Generally speaking, aquarists treat their fish at two different times. The first time is before introducing a new fish to the aquarium, and the second time is when a possible parasite infection has been detected. When it comes to preventative treatment, the treatment will normally take place in a separate quarantine aquarium in order to avoid treating healthy fish. Before starting the treatment, always ask yourself if it is really necessary or if simply keeping the new fish in the quarantine aquarium and looking out for any symptoms of ill health would be enough to safeguard against parasites.

In some situations, it will also be a good idea to move a wounded fish to a quarantine aquarium since a wounded fish can serve as a breeding ground for commonly occurring parasites that would normally not be able to affect the healthy fish in your aquarium. You can have parasites lurking around in your aquarium for years without noticing it, until a wounded fish gives them a chance to multiply and start attacking healthy fish in huge numbers. As you can see, the use of a quarantine aquarium can make parasite treatment unnecessary in many situations. This is a good thing since most parasite treatments are stressful for the fish.

Manual Removal of External Parasites and Gill Parasites

Some parasites that infest the skin and/or gills of the fish can be removed using forceps. Argulus, leeches, copepods, and isopods are all examples of such parasites.

Start by anesthetize your fish using Tricaine Methanesulfonate anesthetic (commonly sold under the name MS-222). Slowly add Tricaine Methanesulfonate to a minimal amount of water with the right salinity to anesthetize the fish. Use forceps to remove external/ gill parasites that are large enough. If your fish is infested with gill isopods you will have to cut the legs of the isopods using small scissors and then use forceps to gently pull the legs from the gills.

Keep a watchful eye on the gills to make sure they continue to work while you work on your fish.

Make the process short and efficient, a fish that is kept in Tricaine Methanesulfonate for any longer period of time will die.

Dipping Away External Parasites

Generally speaking, saltwater dips will remove a lot of external parasites from freshwater species and freshwater dips will do the same for saltwater species. The exact length and frequency of the dips will vary depending on parasite species and fish species. In many cases, three dips that last for five minutes each will be enough. Dipping is stressful for the fish and must be used with caution. Read articles about specific parasites for more detailed instructions.

Anti-Parasite Medication

In order to pick the right medication, you must first identify the culprit. There is no quick fix that will work for all possible fish parasites in the world, even though there are some remedies that are effective against a long row of parasites.

Always double-check the recommendations from the manufacturer of the medication by doing an online search. Some manufacturers want you to use a little too much medication. There is often a fine line between killing the parasites and killing the fish. Experienced aquarists can help you out with the right dosing. Also keep in mind that you particular fish species might be especially sensitive to the active ingredient.

A Not About Filters

Keep in mind that a lot of medications capable of killing microorganisms will kill off the beneficial bacteria in your aquarium. Take appropriate measures to avoid spikes in organic waste and rebuild your biological filter as soon as possible. You can read more about biological filtration in the article about cycling. An increase of organic waste is very dangerous for fish that are already weakened by parasites and a tough treatment.

Chapter 5

Preventing Fish Diseases

Providing the right water chemistry is one of the most important things you can do in order to prevent disease in the aquarium. Having a number of fish in a closed aquarium requires some amount of hard work and only a responsible aquarist can keep his or her fish healthy. Here is a list of water properties that need to be checked as often as possible:

- The temperature of the water has to be kept stable.
- The pH of the water should be at level that is most suited for the species you are keeping in the aquarium.
- Nitrite, Nitrate, Ammonia, Carbon Dioxide and Oxygen levels need to be monitored regularly. These are the various elements that can tip the balance in your aquarium.
- Filter out as much waste as possible. A completely waste free environment is a dream, but getting as close to it as is humanly possible should be the endeavor.
- Some traces of algae are healthy for the fish and are also indications of a healthy habitat within the aquarium.
- Regular water changes are a MUST for any disease-free aquarium
- The Nitrogen Cycle is another important step that many impatient beginners loath to perform.

Strong and healthy fish are typically very resilient toward disease attacks. It is only when the fish are weakened by stress or by factors like persistent aggression that they give in to sickness. Unfortunately, there are times when even the best care falls short of the required care. When fish do fall sick, the best chance for easy and fast cure is when you can catch the sickness in its earliest stages. If you observe your fish regularly, this is not a difficult task at all. Sick fish will start exhibiting strange patterns in their behaviour. Many times, it is when the initial signs have been ignored that the fish really become too ill and reach a sad end.

Sick fish will almost always display at least one of these symptoms:

- Clamped Fins - fins that are held very close to the body and not spread out
- Head standing - Fish swims with its head down.
- Tail Standing - Fish swims with its tail down.
- Lethargy - Fish seems to be tired and stays put in one corner of the aquarium.
- Listing - Leaning on to one side or the other.
- Scales pointing out - The scales of the fish stick out, almost like on a pine cone.
- Red sores on the body.
- White sores and/or spots, or thread like marks on the body.
- An anus that is red or protruding from the body.

Quarantine Aquarium

It is very advisable to have a simple quarantine aquarium. This is the aquarium in which you will first keep your new fish. You can use this aquarium to isolate and cure diseased fish too. In such a case, it is called a hospital aquarium. This is the aquarium in which the fish will recuperate. A small aquarium is all that you need. There is no need to put in any substrate. This will make the aquarium easy to manage. The quarantine aquarium also does not need any plants. A planted aquarium is more difficult to look after. Since there are not plants, you do not need to use a lighting system. The fish will feel more secure in a subdued and dimly lit aquarium.

Filtration can also be kept simple. Chemical filtration, especially if it uses carbon, can deprive your aquarium of any medications that you put inside the aquarium and should therefore not be used. Using a power filter may disturb the ailing fish due to the water disturbance it can cause. A single sponge filter that was kept running in the main aquarium can be used in this aquarium. This will also transport the beneficial bacteria into the new aquarium. It is advisable to put some kind of scenery at the back of the aquarium as this makes the fish feel more secure and protected.

Keep all decorations to the barest minimum, though providing some simple hiding places is advisable. A 50-100W heater will suffice for a small aquarium. Provide an air stone that will maintain high oxygen levels in the water since some medications deplete the oxygen content in water.

Once the fish are out of the hospital aquarium, the aquarium must be cleaned thoroughly before it can be used again. Disinfect the aquarium using Potassium Permanganate or Diluted Hydrochloric Acid. Leave the chemical in the aquarium for a couple of days, and then clean and rinse the aquarium thoroughly. The aquarium is then ready to be setup again when needed.

Red Pest Disease

Red pest is a disease that affects fish and that can cause bloody streaks on the fins, tail and body of the fish. Blood streaks do not have to appear in all these areas for it to be red pest; it will sometimes only affect the body or fins. The disease can in severe cases cause the fins and the tail to fall of. Red pest disease usually infects already weakened fish and is uncommon in healthy fish in well kept aquariums.

Red pest is an internal fish disease and external medications are therefore seldom effective. External remedies can sometimes be effective if the disease is caught very early but it is still recommended to use internal medication instead. The best way to treat red pest is to add antibiotic to the fish food.

Mix antibiotics with flake food carefully. Adding about 1% antibiotics to the food is ideal but a slightly higher dosage can be required in severe cases. Avoid using stronger doses if the case isn't severe as it can have unwanted side effects. It can be good to starve the fish a little before feeding them the medicated flake food to make sure that they eat it. There are a variety of different antibiotics that can be used, e.g. tetracycline and chloromycetin (chloramphenicol). Antibiotics can be added to other foods like frozen food as well in the same concentration. If you for some reason can't get the fish to ingest the antibiotics you can try to add no more than 10 mg of antibiotics per litre water. This will however not be as effective and might not cure the disease.

It is also recommended to clean the aquarium and adding a fish safe disinfectant such as acriflavine (trypaflavine) or monacrin (monoaminoacridine) to the aquarium if red pest is present. Use 1 ml disinfectant per litre water. Stop using the disinfectant if the fish seem to experience discomfort by its presence. The abovementioned disinfectants will colour the water. Disinfectants can be enough to treat the fish in mild cases of red pest but I strongly recommend that you don't rely on that.

Swim-Bladder Disease

Fish use the swim bladder to regulate buoyancy in the water, i.e. adjusting their relative weight to float without sinking or rising in the water. If the fish somehow gets an injury or disease to its swim bladder, regulating its buoyancy will be hard or impossible for the fish. A swim-bladder problem can therefore make the fish swim unnaturally and experience trouble keeping its buoyancy regulated. This can result in the fish laying on the bottom of the tank or floating on the surface. In less severe cases it can just seem like if the fish was drunk and having trouble keeping its balance. Swim bladder disease can be caused by a number of different causes but the most common causes are:

External Factors: If the fish is subjected to external trauma it can affect their swim bladder. External trauma includes physical trauma such as being beaten by another fish, as well as rapid changes in the water parameters in their tank. A rapid temperature increase or decrease can as an example cause swim bladder disease.

Genetic Factors: Some fish can be genetically predisposed to develop a deformed swim bladder when they grow older. This problem is usually a by product of line breeding by breeders trying to breed a certain trait in the fish such as long fins or certain colours. The predisposition is unwittingly being line breed along with the desired traits.

Cancer and TBC: Cancer and tuberculosis can cause swim bladder problems if they affect organs close to the swim bladder or the swim bladder itself.

Diet: Dietary problems such as poor nutrition can cause swim bladder disease as the fish don't get the nutrients they need. An improved diet can often help the fish recover, unless the problem has progressed too far. Poor diet can also lead to constipation which in turn can lead to swim bladder problem as the swelling of the abdomen prevent the swim bladder from working properly.

Diseases: Swim bladder disease can be a by-product of a number of different bacterial and parasitical infections in the fish.

The problem can usually be resolved by treating and resolving the problems listed above. Damages caused by external factors might be permanent and nothing to do about except kill the fish if it doesn't recuperate within a couple of weeks. You should also make sure that you give the fish their preferred environment if you discover swim bladder problems.

Tail and Fin Rot

Tail and fin root is a fish disease caused by bacteria. Normally, this disease will not infect healthy fish living in good conditions, but poor health and poorly kept aquariums are not the only causes of tail and fin root. It can for instance be caused by injuries and other fish nipping the fins or just generally bullying them; all which weakens the fish in a way that allows tail and fin rot to infect them.

This means that it isn't very hard to prevent or at least minimize the risk of tail or fin rot in your aquarium. All you have to do is to keep the aquarium clean and well kept; the fish healthy and well fed, and only combine fish that gets along well. Tail and fin rot can also be caused by Tuberculosis which can be harder to prevent. It is important to determine the cause of the disease when treating it and rectify the problem otherwise the disease might return within a few weeks.

The main symptom of fin and tail rot is exactly what the name suggests; that the fins or tail starts to rot and disintegrate. In severe cases there will be nothing more than stumps left of them. The disintegration of the fins often leaves exposed fins rays and the disease can also cause an irritated area around the base of the find and bloody edges on the fins. Tail and fin rot can also generate symptoms on other part of the fish body such as skin ulcers, loss of colour and cloudy eyes. The skin ulcers usually have red or gray edges.

External treatment by antibiotics is usually enough to treat this disease but the disease can be internally medicated as well. Which option you choose might depend on the size of your aquarium etc as medicating the water in a large tank can be expensive. If you choose to internally medicate your fish you should mix 1% antibiotics (chloromycetin /chloramphenicol, tetracycline or other similar antibiotics) in the food. Make sure that the fish eat the medicated food. If you decide to medicate the water you should add 20-30 mg per litre water of the same antibiotics I mentioned above for internal treatment.

Treat Argulus

To treat and rid your tank of Argulus, commonly known as fish lice, there are several methods one can try. One of the most effective treatments against the fish lice is using an organophosphate.

Organophosphate has been used successfully on removing lice, fleas and other parasites from dogs, cats, horses and even gardens. However, the United Kingdom has banned the use of organophosphates for use a fish disease treatment. You can still purchase it but it will be costly.

Using organophosphate in three treatments over the estimated life cycle of the parasites almost always removes the fish lice. You would do the three treatments at 10 day intervals, which would not only remove the adults and juveniles but any juveniles that may be hatching or recently hatched making sure the fish disease doesn't come right back. A more common organophosphate is known as Fluke Tabs.

Fluke tab treatments should not be continuous, as many types of fish, but especially the larger cyprinids (Koi and Goldfish) may experience a strange inflammatory dermatitis if the product is left in the water longer than 48 hours. Therefore, it is recommended that you do a fifty percent water change after 48 hours, and re-treat after a 24 hour hiatus in the middle.

Organophosphates certainly do eliminate monogenetic trematodes with varying effectiveness, but it is known that these Monogeneans can and DO develop resistance to the compounds. The unique, carbamate-pesticide portion of this compound is the important adjunct that enhances its parasiticidal effect.

Do not use on tropical fish from the Amazon River. Discuses are particularly sensitive. There is significant toxicity of the Carbamates to certain fish, most notably Discus and Catfish, who may be affected adversely.

There is some that recommend using a chitin inhibitor, such as dimilin against this fish disease, which will prevent the juveniles from developing into adults as they shed their exoskeleton but we have not found any conclusive testing to back this theory. Dimilin is said to interrupt the parasite's life cycle and reproductive processes. It is said to be safe for any water plants you may have in the tank and will not discolour the water.

Dimilin should be used as soon as parasites are observed on the fish. Fish lice infestations are easily spread to other tanks by nets boots, clothing & hands. Moving an infested fish to another tank will spread the disease to other fish.

You can also use Copper Sulfate Pentahydrate. Copper Sulfate Pentahydrate is an odorless blue crystal that readily dissolves in water. It is also soluble in methanol, glycerol and slightly soluble in ethanol. Copper Sulfate, sometimes referred to as Blue Vitriol or Bluestone, is a readily available and economic source of copper. You would use one drop per every gallon of water in your tank. Never use this if your water hardness is less than 50ppm, the optimum is 200ppm.

Treating White Spot Disease

White Spot Disease, also known as Ich or Ick, is a very common problem in aquariums and several different methods of treating it has been developed. In this article, we will take a close look at salt and malachite green. If salt or malachite green is the best available treatment for your particular aquarium depends on several different factors, including which species you keep and how sensitive they are to salt. Salt as well as malachite treatment is often combined with an increased water temperature and added aeration.

Malachite Green Treatment

Malachite green must be accurately dosed since it is a powerful chemical that will kill you fish if you overdo it. Some species are less tolerant to malachite green than others and it is therefore a good idea to research your species to find out how much malachite green they can handle. Many manufacturers recommend one drop of their product for each gallon of water in the aquarium to treat White Spot Disease, but a lot of producers sell really concentrated malachite green where one drop/gallon will not be safe to use on sensitive fish species.

A dose of 0.05 milligrams malachite green per litre water is generally considered safe when used for three successive treatments (one every other day) but calculating it can be a bit tricky since different bottles contain different concentrations of malachite green.

Look for the concentration listed on the bottle of the medication. It will normally be in percent, so you need to convert it into milligrams per millilitre. 1 percent equals 1 gram per 100 millilitres of water. A 0.5% malachite green solution will for instance contain 5 mg/millilitre. The recommended dose written on the bottle is often 1 drop per gallon (in the U.S.), and since 1 millilitre is about 15 drops, such a dose will create a concentration much higher than 0.05 milligrams per litre. This is why many aquarists have such bad experiences from malachite green; if you follow the dose recommended by the manufacturer you will often overdose and harm sensitive species. If you purchase a 0,75% malachite green solution, add no more than 4 drops per 10 gallons of water.

Marine Salt Treatment

The Ich parasite does no like salt and using aquarium or marine salt is therefore a good way of getting rid of it, provided of course that you keep fish species that can tolerate the increased salinity. The high salinity should be maintained for two weeks to be sure that all parasites are really gone. Bringing the salinity up to 5 ppt (parts per thousand) will kill nearly always kill the parasites, but may be too strong for salt intolerant fish. If you keep catfish or equally sensitive species, bring the salinity no higher than 2 ppt and hope for the best.

Temperature and Aeration

The use of malachite green or salt is often combined with an increased water temperature when combating White Spot Disease, sine this speeds up the life cycle of the parasite. If you decide to boost up the temperature you should ideally also provide additional aeration since cool water holds oxygen better than warmer water. The increased water temperature will simultaneously increase the metabolism of your fish, which in turn increases their need for oxygen. Since Ich parasites frequently attached themselves to the gills of fish and destroys them, it is easy to understand that an increased water temperature can be very difficult for the diseased fish to handle. Adding additional aeration and keeping the oxygen levels really high will aid them in their struggle.

Treating Fish

There are several ways in which fish can be treated. Unlike other pets, visible changes are not very quick and you can never be sure whether your fish are taking a turn for the better or not. Compared to many other pets, fish are also very delicate. They cannot be taken out, treated and put back into their homes without exposing them to great risks. So, how does one apply medications, feed medicines or help the fish to recuperate?

Giving the fish a bath is a good way to treat some disorders. The fish will be introduced into a bucket containing medicated water. You should use chemicals like formalin or additives like salt in the water you bade the fish in, depending upon what you are treating. The duration of the bath and the concentration of the additive are species-dependent and also depends on the disease. Always follow instructions that are based on the need of your fish species and the disease in question.

Heat therapy is used when it is necessary to kill certain parasites by raising the temperature. One has to be careful while doing this because some fish get too stressed when there is a change in temperature. The temperature of the water should be increased gradually by no more than 1 Degree C per hour until you reach the desired range. The fish should be constantly monitored while the treatment is going on. Any negative reactions call for an immediate cessation of the therapy.

Medicated food products are another effective way of fighting parasites, especially if the disease is inside the body. These medicated foods are not governed or monitored by any governmental agency, so the food concerned may never have been tested for effectiveness or for safety. Therefore, the medicated food that you give your fish may be practically useless or even harmful to some fish. You should purchase only such food that has a detailed list of the ingredients along with their concentration printed on it. Some well known brands can be trusted to deliver quality products, but lesser known brands need to be tested with caution. Awareness of the use and effect of the various medicines is a must before you can start using this kind of treatments.

There are certain steps that should be followed when treating fish. If these steps are followed one after the other, they eliminate not only the symptoms but also the cause of the disease.

The first thing to do when you need to diagnose a fish is to observe it. Observe what signs of disease that are seen and what behavioural patterns have changed, if any. Next, go on to eliminating the possible causes of the signs you se. Check if all environmental conditions are in proper order. The first and the most widespread cause of ailment, disease, stress and death in aquaria is bad water. The ammonia, nitrate, nitrite, pH and oxygen levels have to be checked. If these are within the optimum ranges, the next step is to check whether any toxic substances have dripped into the water by mistake. Paint fumes, cleaning agents, pesticides etc are some things that can get into water quite inadvertently. If the problem is the water quality, changing the water frequently and in smaller quantities will help. If none of these seem to be the cause, then move to the next step.

Now it is time to check the various accessories in the aquarium. Filters, heaters, air pumps etc have to be checked carefully to se that they are working correctly. You should also check if any aggressor has attacked the fish and if that may be the cause behind the symptoms. If you find a problem, fix it immediately.

The fish need to be treated once you have identified the disease. Start by moving the infected fish out of the community aquarium and into the quarantine aquarium. Make sure that your quarantine/hospital aquarium is ready to receive patient/s. The water in the quarantine/ hospital aquarium should be changed daily. About 30-50% of the water should preferable be changed each day. While changing water, care must be taken to add more medicine if required. Also remember that some medicines will remove oxygen from water and aeration is therefore a must in a hospital aquarium. After one day, it is time to move the fish to a shallow bowl or aquarium. Closely examine the fish for any parasites or new signs of infection. Treat the water accordingly. Return the fish to the hospital aquarium once the inspection is completed. This process should be repeated if you are unsure of the proper diagnosis, but not more often than once a day since it is very stressful for the fish. The fish should be kept in isolation for about 10 days even if the fish is showing signs of improvement. This will remove all traces of infection and leave the fish germ free by the time it returns to the community aquarium. This ensures that the spread of infection is curtailed.

While medicating the fish, care must be taken to follow the instructions given. Too low a dose will be ineffective, while higher doses could damage the fish further or even kill it. Carbon, and all absorbing filter media, along with UV sterilizers and ozonizers should be removed before using medications since these will render the medicine ineffective. Never mix two medications in the same aquarium unless there are specific instructions to do so. Once the use of medicines is over, make sure that there are no traces of the medicine left in your aquarium.

In spite of the best of care, if your fish still ails, it is perhaps better to put it out of its misery. This is sometimes the most humane thing that can be done. Severing the spinal cord quickly with a very sharp knife is one of the most painless ways to put a large fish out of its misery. Smaller fish can be dropped into boiling water for a quick death.

Many times your fish will recover when you take care of the water chemistry. It is also important to introduce new fish carefully into your aquarium. New fish are sometimes carriers of harmful parasites. That is why the importance of the quarantine aquarium never can be overstressed.

Do not stuff your aquarium with too many fish. Aquarium overload is another factor that can lead to disaster. The risks of infection are high in an overloaded aquarium. The chances of the infection spreading to other fish are also very high.

Keeping an aquarium is very easy if you understand the various principles behind it and approach it scientifically. There are certain procedures that need to be followed at every stage. When you adhere closely to the beaten path and try to give your fish the best possible conditions, your fish will thrive. Once you have gained more experience and gotten to know your particular specimens better, you can work more freely and improvise in order to suit their particular needs.

Tumors in Fish

There are numerous reasons for tumors in fish and tumors are by themselves not to be considered as diseases or even negative even though they can have negative esthetical effects on the fish that gets them. This does however not mean that a tumor can't be a problem; they can be signs of very dangerous diseases. It just means that tumors aren't necessary a problem or a sign thereof. Tumours can for instance be caused by the thing most of us think about when we hear the word tumour – cancer. Cancer aren't common in most tropical aquarium fish as they are relatively short lived and cancer only rarely have time to develop and become a problem before the fish dies of age. This does however not mean that it never happens. Cancer is more likely and can be a bigger problem in more long living fish species. Sharks don't get cancer and you can therefore rule out cancer if one of your (true) sharks develops a tumor. Some researchers actually hope that sharks will reveal the secret to preventing and curing cancer in humans as well.

Tumors in fish can also be due to virus that has infected the fish. The bulk of all tumors found in fish are however result of genetic traits in that fish and hence not dangerous or a sign of problem. It is possible to surgical remove tumors that are unattractive or that causes problems such for the fish, such as problems eating. You will need a hospital tank for recovery, a scalpel, antibiotics and anaesthesia if you want to try this. I will however not write more about it in this text and I recommend that you obtain more information before trying it, or even better – hire someone with experience to do it for you. I would also recommend that you check the local rules surrounding the keeping of fish and animal cruelty to see if you are allowed to perform such operations yourself in the area you live in.

Uronema Marinum

Uronema marinum is a saltwater parasite that can prove fatal for marine fish. The parasites behind the disease are ciliated protozoa that live freely in marine environments. This disease will as most other diseases only infect already weakened fish and the best way to prevent this disease is therefore to keep a clean aquarium with healthy and well fed fish and avoid overcrowding. You should also make sure to keep new fish in a quarantine tank where you nurture them into good health before letting them into your general population. Uronema marinum usually feeds on bacteria but will attack weak fish as well, eating at its internal organs and muscles. Uronema marinum is often misdiagnosed as it shares symptoms with a number of other bacterial and parasitical diseases. It is however important to diagnose the disease correctly as it is fatal for many types of marine fish. The symptoms of Uronema marinum includes rapid breathing, flashing and loss of colour (the fish becomes pale) as well as weight loss and dehydration.

Fortunately there are several ways to treat this disease. One way is to bath your fish in a freshwater tank for 2-3 minutes (until the fish show signs of stress but no more than 3 min). A variety of this treatment is to place the fish in a hospital tank with low salinity

(1.011) for 21 days. This treatment should not be used with sensitive species. Another method is to use chemical treatments such as Methylene blue, Copper Sulfate or Malachite green. Always follow instructions provided with the product you purchased if you choose to use a chemical treatment. There are also other medications such as the antibiotic Nitrofurazone that will help treat *Uronema marinum*.

Velvet

Velvet is a disease caused by protozoa that infects the fins and body of the fish. The disease is not that uncommon but can easily be treated once diagnosed. Some fish species such as Danios are more susceptible to velvet then others, and once infected they can infect other fish species in the aquarium that do not attract this disease as easily. Species that attract this disease easily are usually more resilient toward its negative health effects and can be infected without becoming seriously ill or die. The fish should regardless of this be treated, since an infect fish as earlier mentioned can infect other fish in the aquarium. The disease is extremely contagious and is usually fatal unless treated.

The main symptom of velvet and what has given it its name is the presence of golden or brownish dust on the body and fins of the infected fish. The disease can in fact look rather good on some fish. The golden dust often first appears around the gills as that often are the first place the disease infects. Other symptoms include clamping of the fins, shortage of breath, scratching on the décor, and infected gills.

Velvet can be treated with Acriflavine (trypaflavine) or copper. Both substances have potentially serious side effects and it is therefore important not to overdose and to gradually change the water in the aquarium once the treatment is completed. Copper can poison and kill a fish while Acriflavine (trypaflavine) can cause infertility in fish. Use 0.2 mg copper per litre aquarium water or 1 ml Acriflavine (trypaflavine) per litre aquarium water. There are some information that increased temperature 80F + /27C + might help cure this disease but this has not been confirmed and using copper or Acriflavine (trypaflavine) is recommended.

The Role of Stress in Fish Disease

Stress: Physiological stress and physical injury are the primary contributing factors of fish disease and mortality in aquaculture.

Stress is defined as physical or chemical factors that cause bodily reactions that may contribute to disease and death.

Many potential fish disease pathogens are continually present in the water, soil, air, or fish. In nature fish are often resistant to these pathogens, and they are able to seek the best living conditions available. Food fish reared under commercial aquaculture conditions are confined to the production unit and are weakened by stress conditions including:

increased fish density and poor water quality (i.e., low dissolved oxygen, undesirable temperature or pH, increased levels of carbon dioxide, ammonia, nitrite, hydrogen sulfide, organic matter in the water);

- injury during handling (i.e., capture, sorting, shipping);
- inadequate nutrition;
- and poor sanitation.

These conditions can result in decreased resistance by the fish, resulting in the spread of disease and parasite infestation.

Stress and injury initially trigger an alarm reaction (fight or flight response), which results in a series of changes within the fish. A blood sugar increase occurs in response to hormone secretion from the adrenal gland as liver glycogen is metabolized. This produces a burst of energy which prepares the animal for an emergency situation.

In addition, the inflammatory response, a defence used by fish against invading disease organisms, is suppressed by hormones released from the adrenal gland. Water balance in the fish (osmoregulation) is disrupted due to changes in the metabolism of minerals. Under these circumstances, freshwater fish absorb excessive amounts of water from the environment (over-hydrate); saltwater fish lose water to the environment (dehydrate), This disruption increases energy requirements for osmoregulation.

Respiration increases, blood pressure increases, and reserve red blood cells are released into the blood stream. Fish are able to adapt to stress for a period of time; they may look and act normal. However, energy reserves are eventually depleted and hormone imbalance occurs, suppressing their immune system and increasing their susceptibility to infectious diseases.

Chapter 6

Internal Structure: Vascular and Neural

Vascular

Blood enters the gills through afferent branchial arteries (ABAs), which receive the entire cardiac output via the ventral aorta. The blood that flows through an ABA feeds the two hemibranchs of an arch, where the blood is oxygenated at the lamellae of the filaments. Oxygenated blood from the filaments of an arch is collected by an efferent branchial artery (EBA), which directs blood to the dorsal aorta for systemic distribution. Upon closer examination of blood flow through the filaments, two distinct, yet interconnected, circulatory systems are apparent: the arterio-arterial and arteriovenous vasculature. Because of the large variations in the anatomy of these systems among species (especially in the arteriovenous vasculature), it is difficult to make generalisations. Therefore, the description that follows is not intended to be an all-encompassing description of the gill circulatory pathways, but aims to introduce some of the more important anatomical features of the vasculature that relate to functional aspects of the physiological processes discussed later in this review. For more thorough accounts of gill vascular anatomy.

Arterio-arterial Vasculature

The arterio-arterial vasculature is often referred to as the respiratory pathway, because it is responsible for the exchange of gases between a fish's blood and its environment. Because of the general similarities in this circulatory system among elasmobranchs, lampreys, and teleosts, these groups will be described together; the respiratory pathway of hagfishes will be described separately. In the arterio-arterial system, blood from an ABA feeds filaments on the hemibranchs of an arch via afferent filamental arteries (AFAs), which travel along the length of a filament. In elasmobranchs and lampreys, the AFA regularly feeds the corpus cavernosum (CC) or cavernous body, which is an extensive network of interconnected vascular sinuses in the afferent portion of the filament, e.g., the spiny dogfish; little skate, *Raja erinacea*; lesser-spotted dogfish, *Scyliorhinus canicula*; Endeavour dogfish, *Centrophorus scalpratus*; sparsely spotted stingaree, *Urolophus paucimaculatus*; Western shovelnose stingaree, *Trygonoptera mucosa*; and Arctic lamprey, *Lampetra japonica*.

Possible functions of the CC are as a hydrostatic support device for the gill filament, a blood pressure regulator, and a site of erythrocyte phagocytosis.

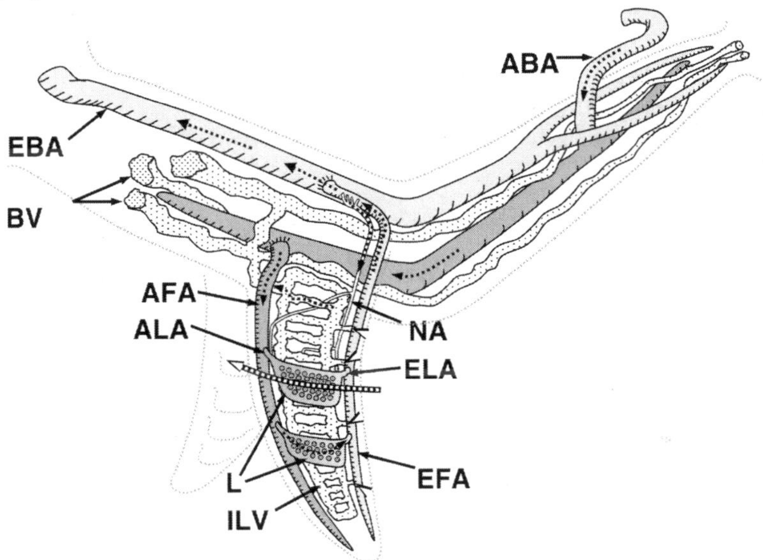

Figure: Generalised schematic of blood flow through the major vessels of a gill arch and filament. *Arterio-arterial pathway*: blood travels (thin dotted arrows) from the afferent branchial artery (ABA) to an afferent filamental artery (AFA), which runs along the length of a filament. The AFA distributes blood to lamellae (L) via afferent lamellar arterioles (ALAs), and the lamellar blood is received by an efferent filamental artery (EFA) via efferent lamellar arterioles (ELAs). Blood flow through the lamellae is countercurrent to water flow across the lamellae (large white-on-black dotted arrow). The EFA returns blood from the filament to the efferent branchial artery (EBA), which distributes blood to the dorsal aorta for systemic circulation. *Arteriovenous pathway*: blood in the EFA can be distributed to the arteriovenous circulation, i.e., interlamellar vessels (ILVs), via postlamellar arteriovenous anastomoses (arrowheads) or by nutrient arteries (NA), which arise from the EFA or EBA. The ILVs are presumably drained by branchial veins (BV), which return blood to the heart.

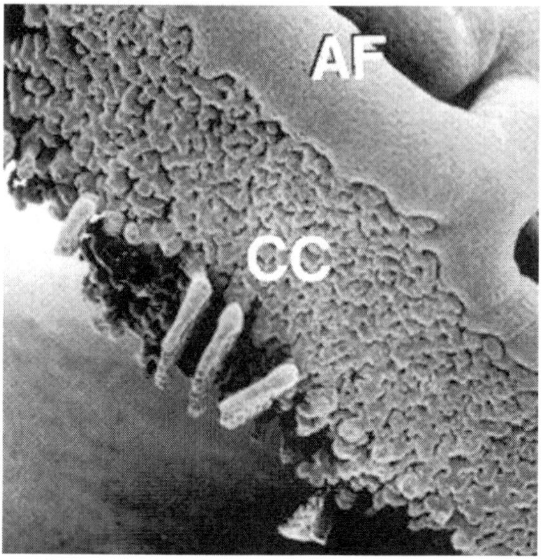

Figure: Scanning electron micrograph (x45) of a vascular cast from the gills of an elasmobranch (Western shovelnose stingaree, *Trygonoptera mucosa*), showing the afferent filamental artery (AF) and corpus cavernosum (CC).

In teleosts, a CC is not present, but in some species the AFAs contain blebs or dilations, e.g., the channel catfish, *Ictalurus punctatus*; European perch, *Perca fluviatilis*; rainbow trout; and skipjack tuna, *Katsuwonus pelamis*. It has been hypothesized that these blebs are vestigial structures derived from the CC of elasmobranchs and agnathans or are functional structures that play a role in dampening pulsatile blood flow.

Regularly spaced along the length of the CC (elasmobranchs and lampreys) or AFA (teleosts) are afferent lamellar arterioles (ALAs) that feed one or more lamella(e). Blood flow through the lamellae is consistent with "sheet flow" theory and occurs through narrow vascular spaces delineated by pillar cells, termed the lamellar sinusoids.

Pillar cells are composed of a central trunk that houses the main body of the cell and thin cytoplasmic extensions (flanges) that spread from the trunk and connect to adjacent pillar cells, thereby lining the lamellar blood space.

In rainbow trout and another teleost, the roach (*Rutilus rutilus*), the lamellar sinusoids lead to deformation, deceleration, and redirection of erythrocytes during their passage through lamellae. This phenomenon is hypothesized to enhance lamellar gas exchange.

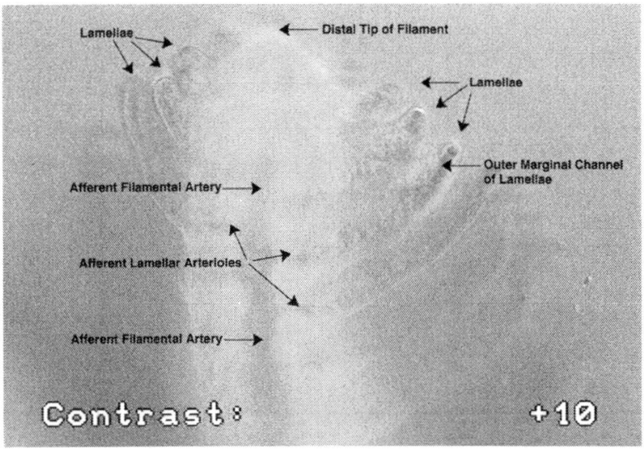

Figure: Video of blood flow through the afferent filamental artery and a series of afferent lamellar arterioles leading into lamellae of a single filament in the gill of the American eel, *Anguilla rostrata*.

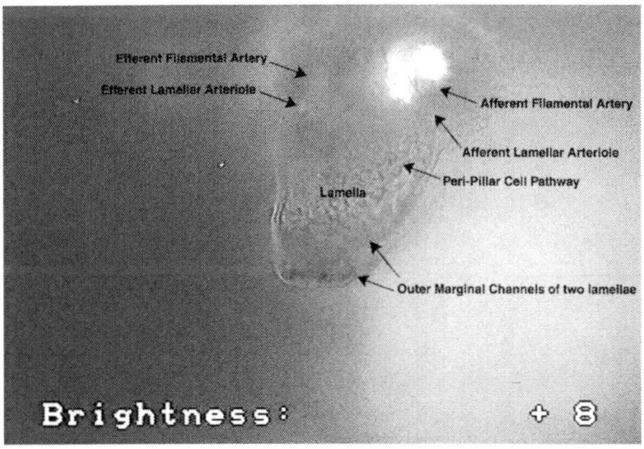

Figure: Video of flow of blood through a single lamella in the gill of the American eel, *Anguilla rostrata*.

Contractile microfilaments are found in pillar cells of several teleosts, lesser-spotted dogfish, and sea lampreys, and evidence for smooth muscle myosin has been detected in pillar cells from the Australian snapper (*Chrysophrys auratus*), a teleost. More recently, FHL5, a novel actin fiber-binding protein, has been localised in pillar cells in the Japanese eel, and its expression increased subsequent to acclimation to fresh water or volume expansion. These findings are suggestive of a contractile nature of pillar cells to possibly regulate blood flow through the lamellae. Recently, in the rainbow trout and Atlantic cod, in vivo videomicroscopy has provided direct evidence for regulation of blood flow through the lamellar sinusoids via pillar cell contraction.

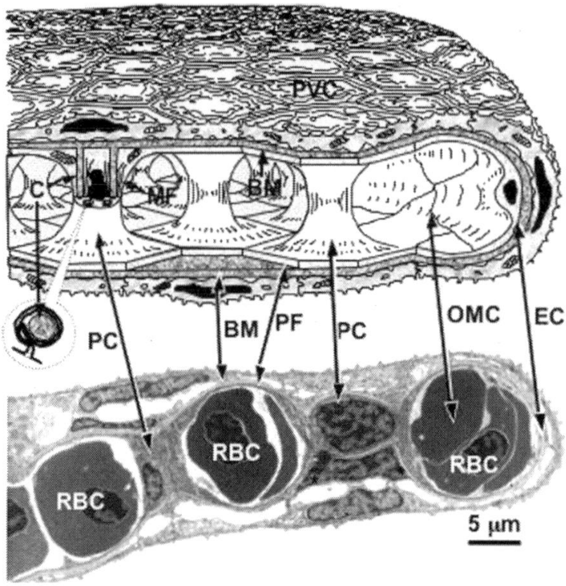

Figure: Schematic and transmission electron micrograph of a cross section through an outer portion of a teleost (rainbow trout) lamella. Note the spool-shaped pillar cells (PC) with their cytoplasmic flanges (PF) that line the lamellar blood spaces, which are filled with red blood cells (RBC) in the micrograph. The pillar cells envelope extracellular bundles of collagen (C), which connect to the basement membrane (BM) that underlies the lamellar epithelium (e.g., PVC). Within the pillar cells are microfilaments (MF) that may be involved with pillar cell contraction. The outer marginal channel (OMC) is formed by pillar cells on its inner boundary and endothelial cells (ECs) on its outer boundary.

In teleosts, elasmobranchs, and lampreys, pillar cells also envelop extracellular collagen bundles, which connect to collagen fibrils of the basement membranes that underlie the lamellar epithelial sheets. The collagen bundles may play a role in anchoring the opposing epithelial sheets to one another, and in maintaining structural integrity of a lamella during increases of blood pressure.

In addition to the lamellar sinusoids, blood flow across the lamellae may take a less convoluted route via the inner marginal channel (IMC) and outer marginal channel (OMC). The IMC is a discontinuous channel at the base of lamellae that is interrupted by pillar cells. This structure is embedded within the body of the filament body, which suggests it does not contribute to gas exchange. Moreover, evidence suggests that the IMC may allow blood to cross lamellae without exchanging gases with the respiratory medium, thus acting as a nonrespiratory shunt. The IMC may also be involved with the collection and distribution of blood within the lamellar sinusoids.

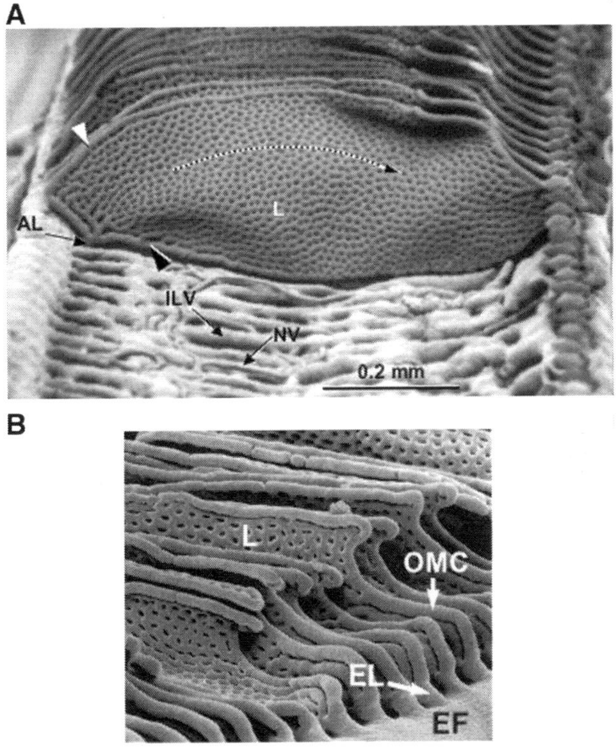

Figure: Scanning electron micrographs of vascular casts from gill filaments of a teleost and an elasmobranch. *A*: a cast of a filament from a teleost (rainbow trout) with some lamellae (L) removed to show the underlying arteriovenous vasculature. Note the afferent lamellar arteriole (AL), inner marginal channel (black arrowhead), outer marginal channel (white arrowhead), an interlamellar vessel (ILV), and a nutrient vessel (NV). Dotted arrow indicates direction of lamellar blood flow. Holes in the lamellae correspond to the position of pillar cells. *B* (x125): a cast of a filament from an elasmobranch (sparsely spotted stingaree, *Urolophus paucimaculatus*) to show the efferent portion of lamellar blood flow. Note the outer marginal channel (OMC), efferent lamellar arteriole (EL), and efferent filamental artery (EF).

In contrast to the IMC, the OMC is a low-resistance, continuous blood space at the periphery of lamellae, lined by endothelial and pillar cells on its outer and inner boundaries, respectively. Compared with the lamellar sinusoids, the OMC has a larger diametre, lower resistance, more erythrocytes, and greater surface area in contact with the ambient water; thus the OMC is hypothesized to be a preferential route of blood flow for shunting large volumes of blood around the lamellar sinusoids to maintain dorsal aortic blood pressure, and for possibly enhancing gas exchange. The latter hypothesis is supported by the recent data that demonstrate that the lamellae of skipjack tuna, a teleost with a high demand for oxygen, have multiple OMCs per lamella and a direct delivery of blood to the OMCs from the ALAs. In rainbow trout and Atlantic cod, the peptide hormone endothelin can redistribute lamellar blood flow to and from the OMC, which may allow these teleosts to alter the functional surface area of their lamellae. This may affect gas exchange as well as the diffusive gains and/or losses of water and osmolytes between a fish's blood and its environment.

Blood is collected from the lamellae by efferent lamellar arterioles (ELAs), which are short vessels that drain into the efferent filamental artery (EFA). The EFA travels along the entire length of a filament's efferent side, and blood flow through this vessel is counter to that of the AFA (i.e., toward the gill arch). The EFA also contains vessels that connect to the arteriovenous vasculature. Near the filament's origin on the arch, the EFA contains a distinctive muscular sphincter before it joins the EBA. This sphincter is highly innervated and may play a role in regulating lamellar blood flow and branchial resistance. The EBA collects oxygenated blood from the filaments and distributes the blood to the dorsal aorta for systemic circulation.

In the Atlantic and Pacific hagfishes, each gill pouch is supplied with blood by an ABA, which connects to an afferent circular artery (ACA) that encircles the excurrent duct of a pouch. Branching off the ACA are afferent radial arteries (ARAs) that feed each gill fold (i.e., filament) in a pouch. An ARA connects to cavernous tissue that permeates the afferent portion of the filament and is comparable to the CC of elasmobranchs and lampreys. At the efferent edge of the cavernous tissue, ALAs arise that provide blood to the highly folded gill lamellae. Interestingly, the complex folding pattern of the hagfish lamellae results in numerous OMCs.

Similar to other fishes, pillar cells within the lamellar folds of hagfish establish the lamellar sinusoids, and, like other fishes, the pillar cells contain microfilaments. Hagfish pillar cells can exist singly and envelope extracellular collagen bundles in a similar manner to other fishes, or pillar cells can occur in clusters of two or more cells that encircle extracellular collagen bundles. Intercellular junctions connect adjacent pillar cells in the Atlantic hagfish, which may suggest a degree of functional synchronisation between the cells, possibly to influence lamellar perfusion. These intercellular junctions are also found in the pillar cells of lampreys but have not been found in any other fishes.

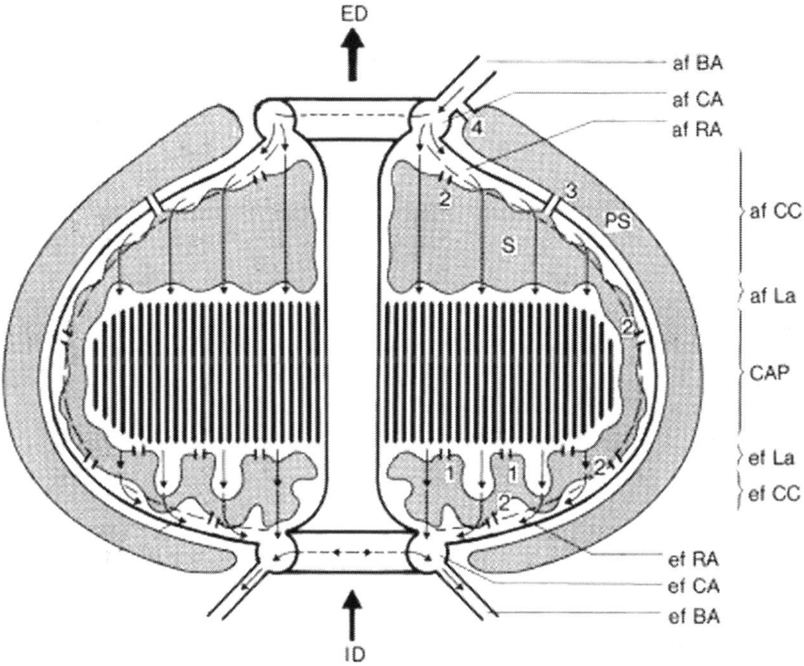

Figure: Schematic of arterio-arterial and arteriovenous vasculature in the gill pouch and filament of hagfishes. The arterio-arterial vasculature includes afferent (af) and efferent (ef) branchial arteries (BA), circular arteries (CA), radial arteries (RA), cavernous tissue (CC), and lamellar arterioles (La). Also noted are the lamellar sinusoids or capillaries (CAP). The arteriovenous vasculature includes the peribranchial sinus (PS) and a sinusoid system (S). The numbers *1*, *2*, and *4* indicate sites of arteriovenous anastomoses. Number *3* marks an anastomosis between the venous sinusoid system and the venous peribranchial sinus. Thin arrows indicate direction of blood flow, and thick arrows indicate direction of water flow.

Blood in the lamellar blood space of hagfishes is drained by ELAs that empty into an efferent cavernous tissue. This cavernous tissue is less complex than its afferent counterpart and feeds an efferent radial artery (ERA). The ERA then joins with an efferent circular artery (ECA) that encircles the pouch's incurrent duct. Blood from the ECA is returned to systemic circulation by two EBAs.

Arteriovenous Vasculature

The arteriovenous vasculature is often referred to as the nonrespiratory pathway. Although the exact function of this system is not known, it is likely involved with providing nutrients to the filament epithelium and the underlying supportive tissues of the gill filaments and may also provide a means for filamental blood to enter the venous circulation without traversing lamellae. The arteriovenous pathway may be considered a component of the systemic circulation because it is primarily supplied with postlamellar blood. Due to general similarities in this circulatory system among elasmobranchs and teleosts, these groups will be described together; the arteriovenous pathway of lampreys and hagfishes are each unique and thus are described separately.

The core of the arteriovenous network in elasmobranchs and teleosts is composed of a highly ordered series of saclike vessels arranged like a ladder. The "rungs" of the ladder are termed the interlamellar vessels (ILVs), which lie underneath the interlamellar epithelium and run parallel to the lamellae. The "legs" of the ladder are termed the collateral vessels or sinuses, which flank and connect to the afferent and efferent boundaries of the ILVs, and run parallel to the length of the filament. The ILVs and their collateral vessels connect to other venous and sinus systems within the filament that return blood to the base of the filament and eventually to the heart.

In the literature, the ILVs are often collectively referred to as the central venous sinus (CVS), because descriptions of the ILVs morphologies vary among fishes and are conflicting. For example, vascular casts by some researchers in teleost and elasmobranchs found that the ILVs primarily appeared to be a continuous, singular, and amorphous sinus below the interlamellar epithelium that ran along the length of the filament, with little secondary structure, e.g., the rainbow trout; European perch; European eel, *Anguilla anguilla*; smooth toadfish, *Tetractenos glaber*; spiny dogfish; and Endeavour dogfish. In contrast, vascular casts by other researchers in some of the same and other teleost and elasmobranch species determined that the ILVs were primarily arranged like rungs on a ladder, e.g., the channel catfish; spiny dogfish; skipjack tuna; walking catfish, *Clarias batrachus*; Asian catfish, *Heteropneustes fossilis*; climbing perch, *Anabas testudineus*; lesser-spotted dogfish; spiny dogfish; little skate; sparsely spotted stingaree; and Western shovelnose stingaree.

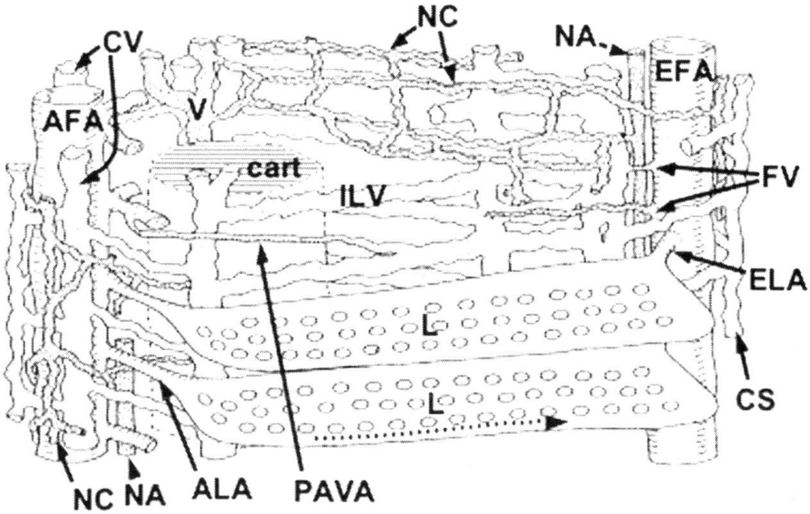

Figure: Schematic of arterio-arterial and arteriovenous vasculature in the gill filament of teleosts. Nonstippled vessels represent the arterio-arterial vasculature, i.e., afferent filamental artery (AFA), afferent lamellar arterioles (ALA), lamella (L), efferent lamellar arterioles (ELA), and efferent filamental artery (EFA). Dotted arrow indicates direction of lamellar blood flow. Stippled vessels represent the arteriovenous vasculature, i.e., interlamellar vessels (ILV), collateral vessels (CV) and sinuses (CS), prelamellar arteriovenous anastomoses (PAVA), postlamellar arteriovenous anastomoses or feeder vessels (FV), nutrient arteries (NA), nutrient capillaries (NC), and filament veins (V). Also note filamental cartilage (cart) that runs along the length of the filament on the afferent edge.

Because the morphology of these vessels in the rainbow trout is sensitive to the perfusion pressure of the vascular casting agent, the latter description is likely the true structure of the ILVs. Therefore, the continuous, amorphous morphology of the ILVs described by some researchers was probably an artifact of superphysiological perfusion pressures, which greatly distend the vessels and create the appearance of a singular sinus. Regardless, the findings of the above-mentioned studies suggest that the ILVs are pliable vessels with an ability to change their structure, and possibly their function, in response to changes in blood pressure.

Blood is supplied to the arteriovenous vasculature by at least three sources: *1*) prelamellar arteriovenous anastomoses (AVAs), *2*) postlamellar AVAs, and *3*) filamental nutrient vessels. Prelamellar AVAs connect the AFA, ALA, or CC to the ILVs. These AVAs are not universal among teleosts and elasmobranchs but have been detected in several species, e.g., channel catfish, smooth toadfish, European eel, Endeavour dogfish, and spiny dogfish. In contrast, postlamellar AVAs (also referred to as feeder vessels) that connect the EFA and/or ELA to the ILVs occur much more frequently than prelamellar AVAs and have been detected in the vast majority of teleost and elasmobranch species examined to date.

It is important to note that blood in the ILVs of teleosts is often characterised by a lower hematocrit than blood found in the arterio-arterial circulation [e.g., the rainbow trout, channel catfish, and black bullhead. These findings suggest that the prelamellar and postlamellar AVAs may be sites of plasma skimming, which emphasizes the nonrespiratory function of the ILVs.

The demonstration of both prelamellar and postlamellar AVAs in some species was consistent with early experimental data that suggested the ILVs may function as a putative lamellar bypass for blood, i.e., a pathway for blood to pass from the AFA (or CC) to the EFA without traversing the lamellar sinusoids. However, this hypothesis was largely refuted, and it is more likely that prelamellar AVAs act as a lamellar shunt, i.e., a pathway for blood to pass from the AFA (or CC) to the ILVs for venous return to the heart without traversing the lamellar sinusoids.

Another source of blood to the ILVs is through nutrient vessels, which compose another meshwork of vasculature within the core of the gill filament termed the nutrient system (NS). It is unclear, however, whether the NS represents a distinct subsystem of the arteriovenous vasculature or is simply another source of blood to the ILVs, like AVAs. Given the infrequent observations of nutrient vessels connecting to the ILVs, it is more likely that the NS is in fact a discrete component of the arteriovenous vasculature that supplies nutrients to the underlying tissues of the filaments and arches and is not just a supplier of blood to the ILVs.

Arterioles and arteries that bud off the EFA and EBA primarily supply blood to the NS. In teleosts, it is common for short, but often tortuous, small-diametre arterioles to branch off the EFA and EBA, and then anastomose into larger diametre nutrient arteries, e.g., rainbow trout, European perch, climbing perch, walking catfish, and skipjack tuna. In elasmobranchs and some teleosts, large-diametre nutrient arteries arise directly from the EFA, e.g., spiny dogfish, little skate, smooth toadfish, and channel catfish. Regardless of how nutrient arteries originate, they travel along the filament's length alongside the EFA or may bifurcate and traverse towards the afferent side of the filament where they branch into nutrient arteries that run medially along the filament's long axis, or into nutrient arteries that run adjacent to the AFA. Additionally, smaller vessels may branch off the nutrient arteries towards the centre of the filament and anastomose with one another to form a weblike arrangement of nutrient arterioles and capillaries that surround, and may anastomose with, the ILVs.

Given the similar location and trajectories of the ILVs and some vessels of the NS, it is often difficult to discern one system from the other with vascular casts, especially in cases when the ILVs are distended and envelop parts of the NS.

However, there are unique morphological features of the vessels that identify these two systems. For example, the vessels of the NS tend to have typical artery/arteriole morphology with a robust endothelium and many erythrocytes, whereas vessels of the ILVs have a more pliable venouslike morphology with a meager endothelium and few erythrocytes. Regardless, both the ILVs and NS are drained by venous systems within the filament that return blood to the base of the filament and eventually to the heart.

In lampreys, the arteriovenous vasculature has not been extensively studied. However, a structure in a similar anatomical location to teleost and elasmobranch ILVs, termed the "axial plate lacunae," has been described in the Arctic lamprey. This structure is openly connected to the CC and EFA without AVAs and appears to function more like an IMC of a teleost lamella than an ILV. The major venous drainage structures in lamprey filaments are the peribranchial sinuses (PBS), which are located at the basal portion of hemibranchs, on both sides of an interbranchial septum. Blood is supplied to the PBS by filament veins, which run parallel to and alongside the AFA. These veins receive blood from the AFA via AVAs that are analogous to prelamellar AVAs of elasmobranchs and teleosts; filament veins then feed the PBS via anastomosing channels along the vein's length. The PBS is also supplied with postlamellar blood via AVAs in the EFA that are analogous to postlamellar AVAs of elasmobranchs and teleosts. From the PBS, blood eventually returns to the heart via other sinuses and veins. It is not known if nutrient vessels are found in the arteriovenous vasculature of lampreys.

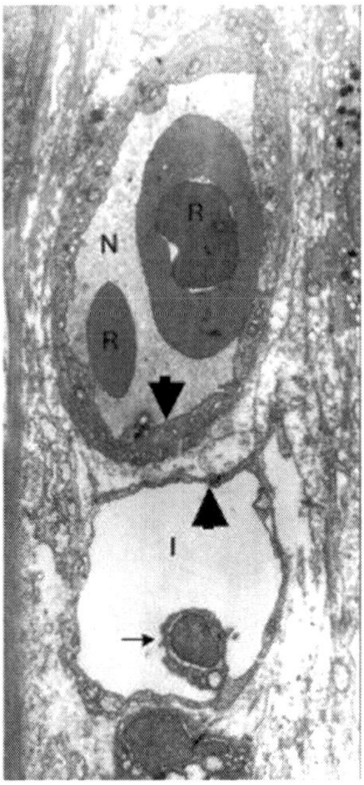

Figure: Transmission electron micrograph (x4,400) of a section through a nutrient (N) and interlamellar (I) vessel. Note the thicker endothelium (large arrows) and occurrence of red blood cells (R) in the nutrient vessel, compared with the interlamellar vessel. Small arrow indicates a nucleus bulging from an endothelial cell.

In hagfishes, the arteriovenous vasculature also has not been extensively studied, but evidence suggests it shares some similarities with that of the otherfish groups. Similar to lampreys, a PBS is present, which surrounds the gill pouch. In the Atlantic hagfish, a sinusoid system, which is analogous to the ILVs of elasmobranchs and teleosts, surrounds the afferent and efferent portions of the filaments, but not the lamellae. In the afferent and efferent cavernous tissues, AVAs supply prelamellar and postlamellar blood, respectively, to the sinusoid system. Other AVAs also have been noted between the sinusoids and the radial vessels, but it is not clear if nutrient vessels are present.

Neural

Branchial nerves in fishes are derived from the VIIth (facial), IXth (glossopharyngeal), and Xth (vagus) cranial nerves, each divided into posttrematic, motor rami, and pre- and posttrematic, sensory rami. In a given gill arch, posttrematic rami from one nerve may mix with pretrematic rami from an adjacent nerve. Gill arches other than the first are generally innervated by the vagus nerve. There are substantial data supporting sensory and motor function for the teleost branchial autonomic innervation; however, similar functional data for elasmobranchs and agnatha have not been published, despite the fact that at least adrenergic and cholinergic receptors have been functionally described in the branchial vasculature of both elasmobranchs and hagfish.

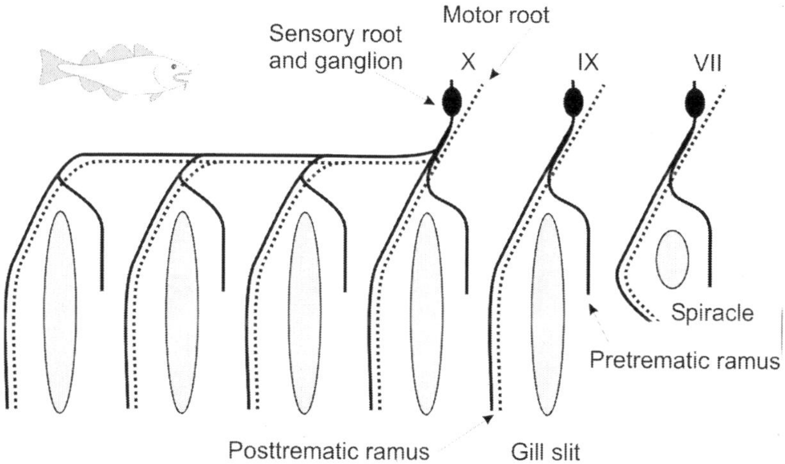

Figure: Outline of the components of the branchial nerves of fish, showing facial (VII), glossopharyngeal (IX), and vagal (X) rami in relation to the gill slits. The orientation is cephalic to the right, as indicated by the outline of the cod.

Histochemical studies have described catecholamine-containing fibres associated with AFAs and ALAs in a variety of teleost species, and similar fibres associated with ELAs and EFAs in some species, but the lamellae do not appear to be innervated. On the other hand, catecholamine-containing fibres have been described around the ILV in a number of teleost species. Cholinergic-type fibres have been described in the gills but are only associated with the efferent vasculature, specifically with a prominent sphincter in the basal third of the efferent filamental artery, which also appears to contain serotonergic fibres. Serotonergic neuroepithelial cells (NECs) have also been described on the gill filament in a variety of species, as have fibres and NECs that contain immunoreactive nitric oxide synthase. Specific roles for the neurosecretory products of these fibres and cells are discussed in sections IV*D* and VIII*B*.

Gas Exchange and Gas Sensing

Because the branchial epithelium is essentially a monolayer of mixed cells (mostly squamous PVCs on the lamellar surface), and the medium irrigates the external surface countercurrent to the internal perfusion, the fish gills appear to be ideal for gas exchange. The ratio of irrigation (ca. 5,000–20,000 ml $\cdot$ kg^{-1} $\cdot$ h^{-1}) to perfusion (Vg / Q) is generally of the order of 1–8 but may approach 50 in very active, pelagic fishes such as tuna (by substantial increases in $_g$), and oxygen extraction efficiencies vary from 50 to 90%, higher than those described for mammals.

The metabolic cost of gill gas exchange is substantial, however, because of the low solubility of oxygen in aqueous solutions (~7 ml/l; 3% that in air) and the high viscosity (800 times air) and density (60 times air) of the medium. Thus the cost of routine ventilation may be 10% of the oxygen uptake (O_2), and it may approach 70% of MO_2 during exercise, when ventilation volumes may triple. Pelagic fish species reduce the cost of ventilation by utilising forward movement and open mouths to drive water across the gills.

Lamellar Gas Exchange

The actual site of gas exchange is assumed to be the lamellar epithelium, but no direct measurements of lamellar versus filamental gas permeabilities have been published. The lamellar surface is likely the primary site, because various studies have shown a correlation between lamellar surface area and respiratory needs (e.g., benthic vs. pelagic) or measured rates of oxygen uptake.

In fact, recruitment of distal lamellae is generally considered to be a means of increasing gas exchange, although this hypothesis is largely based on two studies that demonstrated that only 60–70% of lamellae in rainbow trout and lingcod (*Ophiodon elongatus*) were perfused at rest. In the trout, lamellar perfusion was measured by the distribution of vitally stained erythrocytes after infusion into trout via a cannula in the ventral aorta; in the lingcod, the distribution was measured by perfusion of isolated branchial arches with saline containing Prussian Blue dye.

In both cases, it is unclear if lamellar perfusion mimicked the in vivo condition. Our own videography of blood flow through the gills of American eels (*Anguilla rostrata*) or longhorn sculpin (*Myoxocephalus octodecimspinosus*) indicates that even distal lamellae are perfused in anaesthetised individuals, so lamellar recruitment may not be a general phenomenon. The respiratory (lamellar) surface area is generally 0.1–0.4 m^2/kg but may be as high as 1.3 m^2/kg in tuna (equivalent to human) or as low as <0.1 m^2/kg in species that utilise the swim bladder (or other accessory air-breathing organs) for oxygen uptake. The respiratory medium (water) to blood diffusion distance (i.e., the thickness of lamellar epithelium) also varies with life-style, ranging from 10 μm in aerial breathers to <1 μm in tuna. It is not clear if this dimension can be dynamically regulated, but thinning of the epithelium might possibly occur if an increase in cardiac output distends lamellae. In addition, alteration in the distribution of gill cells (MRC vs. PVC) under hypoxic or hypercapnic conditions may alter the diffusion distance and thereby impact gas exchange.

Perfusion Versus Diffusion Limitations

The uptake of oxygen across the fish gills appears to be perfusion limited, but all data published to date are from a single species, the rainbow trout, so more data are needed to support the general statement. Nevertheless, the fact that rainbow trout arterial oxygen content (Pa_{O2}) did not vary appreciably with changes in perfusion produced by alteration of blood volume argues for perfusion limitation, as does the lack of any alteration in Pa_{O2} subsequent to the ligation of two gill arches. In contrast, excretion of CO_2 by the rainbow trout gill epithelium appears to be diffusion limited, at least in teleosts. Thus alterations in blood volume in the trout were correlated with alterations in blood CO_2, and in rainbow trout a 30% reduction in gill surface area was accompanied by a significant elevation of Pa_{CO2}. For a more complete discussion of the data supporting perfusion limitation for O_2 uptake and diffusion limitation for CO_2 excretion across the fish gills. The diffusion limitation to CO_2 excretion in teleosts is apparently secondary to the need for erythrocyte Cl^-/HCO_3^- exchange (via band 3) at the gills, the only means of CO_2 production at the gas exchanger because of the lack of endothelial carbonic anhydrase (CA), as is found in the mammalian lung alveoli. The recent findings that injection of CA offset the increased Pa_{CO2} seen in the rainbow trout after an increase in cardiac output or decrease in gill surface area support this conclusion. It is unclear if the constraints of CO_2 excretion are the same in most fishes, but it is apparent that the dynamics of CO_2 excretion across the gills in at least three other fish taxa may be different. For instance, elasmobranchs do have vascular CA, hemoglobin-less icefishes lack erythrocytes and therefore band 3, and lampreys are unique among vertebrates in the lack of a functional anion exchanger in the erythrocyte. One might expect CO_2 excretion in these three groups to be perfusion limited, but this hypothesis has not been tested to date. Despite these theoretical constraints on CO_2 excretion, the fish gills appear to be hyperventilated with regard to CO_2, because the blood Pv_{CO2}, and bicarbonate concentration (~4 mM) is far below that found in mammals, producing a plasma pH (~7.8) that is substantially higher than that measured in mammalian plasma. The impact of these values on acid-base regulation is discussed in section VI*B*.

Chemoreceptors

O_2 Sensors

Because the branchial arch arteries of fishes are the evolutionary precursors of the carotid and pumonary arteries in mammals (402), one might suspect that the gills would be the site of afferent sensory neurons whose efferent motor output may control cardiovascular and respiratory responses to environmental and/or systemic changes in dissolved gas concentrations.

This is, in fact, the case. As is found in other vertebrates, changes in environmental or blood gas tensions elicit a suite of ventilatory and cardiovascular responses in fishes. The primacy of O_2as the proximate signal is well established, but recent studies have demonstrated that CO_2/pH also may be a relevant signal. Environmental hypoxia elicits hyperventilation in a variety of species, and the response is generally larger changes in stroke volume than rate, thereby reducing the energetic effects of irrigating the gills with a relatively dense and viscous medium; hyperoxia elicits the opposite response.

The responses occur within seconds of the exposure, suggesting external receptors, and this hypothesis is supported by the finding that externally applied cyanide elicits a hyperventilatory response. The receptors are distributed throughout the gill arches and are innervated by the 9th and/or 10th cranial nerves. There are also internally oriented, oxygen chemoreceptors in the gill vasculature that respond to either hypoxemia (in the absence of external hypoxia) or injection of cyanide. Such receptors would be advantageous during exercise or other changes in metabolic rate. It is not clear whether the external or internal chemoreceptors respond to Pa_{O2} or oxygen content, although it is more likely that tension is the proximate signal since vagus nerve recordings from isolated gill arches from both tuna and rainbow trout showed responses to reductions in perfusate Pa_{O2} even when total O_2 concentration was unchanged. Extrabranchial, buccal O_2-sensitive chemoreceptors have also been described, in particular in air-breathing fishes, but there is no convincing evidence to date for O_2 receptors in the central nervous system of fish.

The neuroendocrine physiology of gill oxygen chemoreceptors is relatively unstudied, although current evidence suggests that neuroepithelial cells (NEC) are involved, as they are in the lungs of terrestrial vertebrates. Dunel-Erb et al. described NEC on the gill filaments (near the efferent filamental artery, facing the afferent respiratory water flow) of a variety of teleosts and one elasmobranch, that degranulated under hypoxic conditions, and serotonergic NEC have been confirmed in other species.

More recently, it has been shown that zebrafish (*Danio rerio*) respond to hypoxia, and the filaments and lamellae of all four gill arches of the developing embryo contain both serotonergic and nonserotonergic NEC. Innervation patterns in this preparation suggest that the serotonergic NEC may be generating signals back to the central nervous system (to modulate respiratory and cardiovascular parametres) as well as releasing paracrines to affect local perfusion patterns within the filament and lamellae. The nonserotonergic NEC are immunopositive for the purinergic , not innervated, and often associated with the pillar cells of the lamellae, suggesting a paracrine role for these NEC.

CO_2 Sensors

The general finding that ventilation is matched to oxygen demands, even at the expense of blood pH (e.g., hyperventilation in hypoxic conditions is accompanied by respiratory alkalosis), suggests that O_2 demand overrides acid-base disturbances. This does not mean, however, that a CO_2/pH ventilatory drive does not exist. For instance, increasing the Pa_{CO2} of a hyperoxic medium abolishes the usual hypoventilatory response in rainbow trout or may even produce hyperventilation in the channel catfish. Indeed, the physiology of putative CO_2/pH-sensitive receptors in the gill epithelium has generated much recent interest. Fishes from all major taxonomic groups respond to environmental hypercapnia by a significant increase in gill irrigation and bradycardia. Until recently, it was assumed that the responses were indirect, the result of hypoxemia produced by the effect of blood hypercapnia on hemoglobin oxygen affinity (Bohr effect) and carrying capacity. More recent studies, however, have demonstrated a direct effect of CO_2/pH on cardiorespiratory parametres, under normoxemic conditions.

The CO_2/pH-sensitive chemoreceptors appear to be primarily on the first gill arch in some species but may be more generally distributed in other species, because the responses are abolished only when all gill arches are denervated. The receptors appear to respond to external rather than internal changes in CO_2/pH, because injection of acetazolamide (which raises blood Pv_{CO2}) did not alter ventilatory frequency or amplitude in the rainbow trout, catfish, and spiny dogfish, nor did a bolus infusion of CO_2-enriched saline into the posterior caudal vein of the spiny dogfish. On the other hand, the fact that postexercise hyperventilation in the rainbow trout was reduced by injection of CA (which reduced the attending respiratory acidosis) suggests that internal receptors may be present.

In all tetrapod vertebrates, receptors in the central nervous system respond to changes in perfusate CO_2/pH, but it is unclear if such receptors are present in the central nervous system of water-breathing fishes. Forty years ago, an experiment on a single tench (*Tinca tinca*) found that injection of slightly acidic saline into the posterior medulla increased the amplitude of electrical discharges that were correlated with respiratory movements, and a subsequent study demonstrated an inverse correlation between the bicarbonate concentration of the saline perfusing the isolated brain of the lamprey (*Lampetra fluviatilus*) and the frequency of putative respiratory motor output. Recent studies of water-breathing fishes have not corroborated these findings.

For instance, hypercapnia-induced hyperventilation in the skate (*Raja ocellata*) was correlated with the pH of the arterial blood, not the cerebrospinal pH or intracellular pH of brain tissue, and similar findings have now been reported for other species. On the other hand, aerial gas exchange has arisen multiple times in the piscine lineage (e.g., gars, eels, tarpon, lungfishes), and it appears that central CO_2/pH chemoreceptors have evolved in at least some of these groups. For example, reduction of the pH of the solution perfusing the brain stem-spinal cord preparation of the gar (*Lepisosteus osseus*) was correlated with a significant increase in the putative air-breathing motor output, but not the motor output thought to be correlated with gill ventilation.

Similar results have been described for the South American lungfish (*Lepidosiren paradoxa*), where lowering the pH of the perfusate in the IVth ventricle was associated with an increase in lung ventilation frequency. Thus the extant data do not allow us to stipulate exactly when central CO_2/pH receptors evolved in fishes, but most likely it was associated with either the independent evolution of aerial respiration in early fish lineages (represented by gar) or with the lineage that gave rise to the amphibia (represented by lungfishes). Clearly more data are needed on this interesting question.

A fall in intracellular pH is generally considered to be the proximate stimulus in the mammalian carotid chemoreceptor, and it is likely that this is also the intracellular stimulus in the fish chemoreceptor. Recent evidence suggests that the proximate, extracellular stimulus in the fish gill chemoreceptor is actually a change in extracellular CO_2 rather than pH. For instance, changes in medium pH did not elicit the usual cardiorespiratory responses of two tropical teleosts if the Pv_{CO2} was unaltered. In the Atlantic salmon and spiny dogfish, CO_2-enriched water produced the expected bradycardia and hyperventilation, but acidified, CO_2-free water did not elicit cardiorespiratory changes.

In summary, respiratory function in fishes (using the gills) is similar to that in tetrapod vertebrates (using lungs), but medium-constrained differences do exist. There is a need for more studies in a variety of fish groups before some of the generalities presented above can be verified.

Osmoregulation and Ion Balance

The structure and function of the branchial epithelium as a site of gas exchange dictates that it is also the site of osmosis and ionic diffusion if gradients exist. The marine, invertebrate ancestors of the vertebrates (i.e., sister taxa to modern echinoderms and tunicates) were and are isosmotic (but not always isoionic) to seawater. In contrast to an early hypothesis of freshwater origin, it is now generally accepted that the first vertebrates evolved in seawater, entered brackish and fresh water (one lineage giving rise to the Amphibia), and then reentered the marine environment in some cases. The suggestion that the presence of a renal glomerulus in nearly all vertebrates is the result of origin in fresh water is invalidated by the presence of functional glomeruli in the totally marine hagfish, modern members of the earliest fish lineage, which do not have a history of freshwater ancestry. The scenario of marine origin followed by entry into fresh water also is supported by the extant fossil record and is the most parsimonious explanation for the isotonicity of hagfish to seawater and the general finding that the NaCl concentration of the blood and extracellular fluids of all the other fish taxa (indeed, all other vertebrates) is ~30% that found in seawater, but distinctly above the NaCl content of fresh water. The maintenance of blood and intracellular tonicity in the face of these osmotic and ionic gradients across the fish branchial epithelium takes energy, but the most recent calculations suggest that it is only of the order of 2–4% of the resting oxygen consumption, far below the cost of ventilation.

Osmoregulation in Fresh Water

Like all other freshwater animals, modern fishes (and, presumably, their evolutionary precursors as they entered brackish and fresh water) balance the osmotic uptake of water by having substantial glomerular filtration rates and urine flows and minimise renal salt loss by a significant tubular reabsorption of necessary ions. Net ionic losses in the urine, and by diffusional outflux across the gill, is balanced by active uptake mechanisms in the gill epithelium plus any ionic gain from food. For general reviews of freshwater fish osmoregulation.

Apical Na^+ Uptake

The mechanisms of ionic uptake were first suggested by August Krogh, who found that Na^+ and Cl^- were extracted from fresh water independently by the goldfish (*Carassius auratus*) and suggested that blood ions such as NH_4^+, H^+, and HCO_3^- could function as counterions to maintain electroneutrality across the epithelium. Subsequent studies appeared to support this hypothesis, but uptake of Na^+ via passive exchange with either NH_4^+ or H^+ has been questioned on thermodynamic grounds.

For instance, gill epithelium intracellular Na^+ concentrations have been reported in the range of 50–90 mM in two species of trout and ~10 mM in tilapia, far above the common freshwater concentration of <1 mM Na^+. In addition, it has been calculated that a pH gradient of 0.3 units would be necessary to drive H^+ outward (in exchange for external Na^+), but rainbow trout are able to excrete protons into water with a pH of 6 despite the fact that the gill epithelial cell pH in this species is ~7.4, and blood pH is even higher. In addition, many of the early studies that appeared to support the model of Na^+/H^+ or Na^+/NH_4^+ exchange by demonstrating sensitivity of Na^+ uptake to amiloride used concentrations (0.1–1 mM) that could not distinguish among inhibition of ionic exchange, blockade of an apical Na^+channel, or even inhibition of basolateral Na^+-K^+(NH_4^+)-ATPase. Moreover, the Na^+/H^+ exchanger (NHE)-specific inhibitor dimethylamiloride did not inhibit Na^+ uptake by the isolated European flounder (*Platichthys flesus*) gill filament.

Recent data suggest that the uptake of Na^+ by the fish gill epithelium may be electrically coupled to proton efflux, rather than directly coupled by a single exchanger (i.e., NHE) to proton excretion. For instance, it was found that the HCO_3^- concentration in the external water (at a neutral pH) actually fell as the medium flowed past the rainbow trout gills, contrary to what one might expect if excreted CO_2 was hydrated in the unstirred layer on the water side of the gill epithelium. Moreover, the fall in HCO_3^- concentration was not altered by the disulfonic stilbene SITS, so it was probably not due to reversed Cl^-/HCO_3^- exchange. These data suggest that secreted protons dehydrate the HCO_3^- in the unstirred layer.

Addition of vanadate to the external medium at a concentration (0.1 mM) that does not inhibit basolateral Na^+-K^+-ATPase in the frog skin inhibited 50% of the putative proton secretion by the rainbow trout gills, suggesting that the excretion was via an apical, P-type proton pump. Other studies described a vesicular fraction isolated from rainbow trout gills that transported protons and expressed a Mg^{2+}-dependent ATPase that was sensitive to *N*-ethylmaleimide, a known inhibitor of both V-type and P-type H^+-ATPases, but with a 10- to 100-fold selectivity for the former. Moreover, one study was in the presence of EGTA, azide, and ouabain, so activity of Ca^{2+}-activated ATPase, mitochondrial H^+-ATPase, and Na^+-K^+-ATPase was assumed to be minimal. In addition, the NEM-sensitive ATPase activity was also inhibited by other H^+-ATPase inhibitors, such as dicyclohexylcarbodiimide (DCCD), diethylstilbestrol (DES), *p*-chloromercuribenzenesulfonate (PCMBS), and bafilomycin. The inhibitor profile did not allow a definitive differentiation between P- and V-type proton pumps, but subsequent physiological and immunological studies have made a V-type proton pump (V-H^+-ATPase) the most likely. For example, the V-H^+-ATPase specific inhibitor bafilomycin (1–10 μM in the freshwater medium) inhibited 60–90% of Na^+ uptake by tilapia, carp (*Cyprinus carpio*), and zebrafish. These results corroborate molecular studies that showed that heterologous antibodies raised against various subunits of V-H^+-ATPase localised expression to the branchial epithelium of rainbow trout, coho salmon, tilapia, zebrafish, and two elasmobranchs: the spiny dogfish shark and Atlantic stingray.

Homologous antibodies raised against the zebrafishisoforms demonstrated expression in gill tissue from that species. In addition, the mRNA for V-H^+-ATPase has been localised in the branchial epithelium of the rainbow trout by Northern blot, using a probe complementary to the cloned B-subunit of the rainbow trout gene as well as by in situ hybridisation using probes complementary to the bovine renal proton pump. The amino acid sequence for this rainbow trout gill V-H^+-ATPase subunit shares >95% identity with the vatB1 isoform of the eel swim bladder gas gland V-H^+-ATPase and zebrafish gill vatB1, as well as the vatB1 isoform from humans. Similar amino acid sequence identities have been shown for the vatB2 isoform from the European eel gas gland and zebrafishgills compared with the human sequence, but the gene for the vatB2 isoform has not been cloned from the rainbow trout gill. A partial sequence from the - subunit from the gill of the Osorezan dace (*Tribolodon hakonensis*) has been reported, as well as a full-length clone of the á-subunit of V-H^+-ATPase from the killifish gill.

The presence of V-H^+-ATPase in the fish branchial epithelium supports the hypothesis that Na^+ uptake by this tissue may be coupled to electrogenic proton extrusion, as has been described for tight epithelia such as frog skin and distal portions of the mammalian nephron. The multistranded junctional complexes between MRC and PVC cell in the branchial epithelium of freshwater fishes suggest an electrically tight epithelium, and the extremely high resistance of the cultured, freshwater rainbow trout gill epithelium (3.5 kÙ · cm^2) supports this proposition. If the "tight epithelium" model of electrical coupling between proton extrusion and Na^+ uptake is correct, then V-H^+-ATPase and a Na^+ channel should be colocalised to the apical membrane, and Na^+-K^+-ATPase should be localised to the basolateral membrane of the same cell. The initial immunolocalisation of V-H^+-ATPase suggested that both MRC and PVC in the rainbow trout gill expressed the proton pump, and the labelling in both cells now has been shown to be apical.

In the coho salmon, V-H^+-ATPase was localised to MRC. Another study on the rainbow trout, however, has localised the proton pump in the apical region of only PVCs, and this PVC-only localisation has now been extended to tilapia.

This localisation of the proton pump to the PVC was also supported by PVC versus MRC distribution changes under acidotic conditions, high-magnification transmission electron microscopy, and X-ray microanalysis studies. Whether these differences in the cellular localisation of V-H^+-ATPase are antibody specific or due to differences in rainbow trout populations, holding conditions, or species, remains to be seen, but current data do support an apical position for the proton pump in one or more types of epithelial cells in the freshwater fish gill. An amiloride-sensitive Na^+ channel has not been cloned from fish, but benzamil (an amiloride analogue with a high affinity for ENaC) inhibited Na^+ uptake by freshwater European flounder gills. In addition, a recent study demonstrated that an ENaC polyclonal antibody (raised against a synthetic peptide corresponding to residues 411–420 of the -subunit of the human epithelial Na^+channel) colocalised with the V-H^+-ATPase, in only PVC in tilapia and in both PVC and MRC in the rainbow trout.1 In all of these studies, MRC cells were identified either by their morphology (and/or position in the epithelium) or by immunological staining with heterologous antibodies raised against Na^+-K^+-ATPase, which is assumed to have high expression in the MRC. Recent cell separation techniques may offer an avenue for better delineation of specific cellular function.

It has been demonstrated that PVC and MRC from the rainbow trout gill can be separated by Percoll density gradient, combined with peanut lectin agglutinin (PNA) binding and magnetic bead separation. With the use of this technique, MRC can be separated into two subpopulations, one of which (PNA negative; termed á-MRC) displays bafilomycin-sensitive, acid-activated ^{22}Na uptake, which also is sensitive to the amiloride analogue phenamil, an ENaC inhibitor. Importantly, in this study, the PVC (defined by position in the Percoll gradient) displayed very low rates of Na^+ uptake that were not sensitive to bafilomycin.

This suggests that the acid-secreting cells in the rainbow trout gill are a subpopulation of MRCs, not PVCs. Thus data supporting the tight epithelium model for Na^+ uptake (and H^+ extrusion) exist for the branchial epithelium of freshwater teleosts, but the data are restricted to two species (rainbow trout and tilapia) and must be extended to a wider array of species. This model for Na^+ uptake across the apical membrane of the freshwater fish gill epithelium may apply to Na^+ uptake from very dilute solutions in some teleosts. But passive, chemically coupled Na^+/H^+ exchange could mediate Na^+ uptake if electrochemical gradients for either Na^+ or H^+ are favourable (e.g., in solutions with a $[Na^+]$ higher than a few mM and/or pH >7.6, which may be found in some fresh water).

Under these circumstances, one might propose apical Na^+/H^+ and Cl^-/HCO_3^- exchangers coupled to basolateral Na^+-K^+-ATPase. These are the apical transporters responsible for Na^+ absorption in mammalian intestinal and renal proximal tubule cells and therefore would be expected to be the isoforms if apical NHEs function in freshwater fishes. Antibodies raised against human proteins have been used to localise NHE3 immunoreactivity in the branchial epithelium of freshwater rainbow trout and NHE2 immunoreactivity in the branchial epithelium of freshwater tilapia.

Neither study specifically localised Na^+-K^+-ATPase to differentiate between MRC and PVC, but morphology and position of the stained cells suggested that the NHEs were only in PVC in the rainbow trout and in both PVC and MRC in tilapia. The most convincing evidence that an apical NHE is expressed in gills in freshwaterfishes comes from a recent study that cloned a full-length transcript homologous to NHE3 from the gills of the Osorezan dace and used a homologous antibody to localise expression to the apical surface of MRC.

Data from another species, the killifish, suggest a third alternative for the distribution of the pumps and channels necessary for at least Na^+ uptake by the fish branchial epithelium. V-H^+-ATPase has been localised to the basolateral membrane of cells that also express Na^+-K^+-ATPase, using homologous antibodies (raised against a peptide fragment coded by cloned, killifish V-H^+-ATPase). This suggests that basolateral cation extrusion by both pumps may produce an electrochemical gradient across the apical membrane that is sufficient to drive Na^+ into the cell from the external fresh water.

It is not clear if environmental or species differences can account for these disparate models for Na^+ uptake across the freshwater fish gill epithelium, but it is clear that molecular techniques, applied to a variety of species and conditions, are needed to provide a better delineation of the mechanisms of Na^+ uptake.

Apical Cl^- Uptake

Chloride uptake by the freshwater fish gill generally is considered to be via Cl^-/HCO_3^- exchange, but only a few studies have examined this putative pathway directly. Inhibitors of this anionic exchange reduce Cl^- uptake and produce metabolic alkalosis in freshwater fishes, and polyclonal antibodies raised against rainbow trout AE1 (anion exchanger, isoform 1) localised to the apical surface of gill cells in tilapia and coho salmon that also were immunopositive for Na^+-K^+-ATPase. In addition, an oligonucleotide probe, complementary to rat AE1 cDNA, hybridised to RNA in cells in both the filament and lamellae of the rainbow trout, suggesting that both PVCs and MRCs contain the message for AE1.

Recently, an alternative mode for Cl^-/HCO_3^- exchange across the apical membrane of the elasmobranch gill epithelium has been proposed. With the use of a heterologous antibody to human pendrin (a non-AE1 anion exchanger), pendrin-like immunoreactivity has been localised to the apical region of V-H-ATPase-rich, but not Na^+-K^+-ATPase-rich, cells in the branchial epithelium of the Atlantic stingray and spiny dogfish. This colocalisation is remarkably similar to V-H-ATPase-rich, HCO_3^--secreting intercalated cells in the mammalian collecting duct.

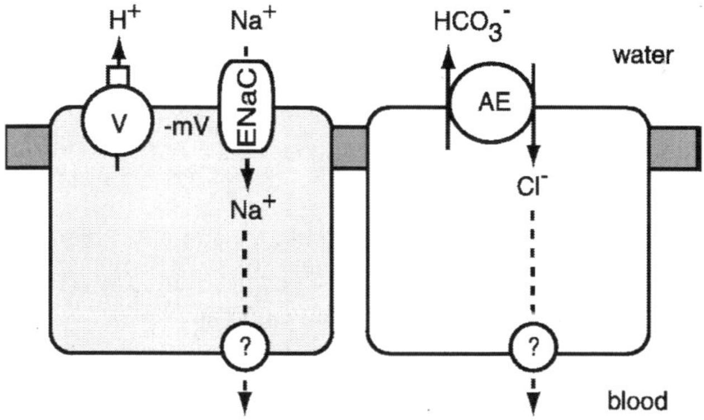

FW Salmonids and Tilapia

Figure: Working model of NaCl uptake mechanisms proposed for freshwater Atlantic stingrays. One type of MRC (A MRC) expresses Na^+-K^+-ATPase on its basolateral membrane and is hypothesized to draw in Na^+ across the apical surface in exchange for cytoplasmic H^+. The other MRC (B MRC) expresses V-H^+-ATPase (V) on its basolateral membrane and draws Cl^- into the cell via pendrin (PDN). The pathway for basolateral Cl^- movement is unknown.

Intracellular CA

The branchial epithelium also contains CA, which is generally thought to be necessary for the production of H^+ and HCO_3^- for the apical exchangers, pumps, and channels that are present. Histological studies (using the cobalt phosphate/cobalt sulfate vital stain) demonstrated the presence of CA in the apical region of both MRCs and PVCs in the branchial epithelia of a variety of fish species and immunohistochemical studies (using homologous antibodies raised against purified enzyme) have confirmed this localisation to the apical region of MRCs and PVCs in both the rainbow trout and European flounder. A CAII-like gene has been sequenced from cDNA from the gills of the Osorezan dace, and the product has been localised to the MRC by homologous immunohistochemistry. In both teleosts and elasmobranchs, this intracellular CA represents the majority of CA activity in gill homogenates. A role for CA in at least Na^+uptake had been proposed, because injection of acetazolamiode into either the rainbow trout or small spotted dogfish shark inhibited gill Na^+ uptake, and it has been demonstrated that exposure of zebrafish to ethoxzolamide (another CA inhibitor) reduced both Na^+ and Cl^- uptake.

Basolateral Na^+ and Cl^- Exit

As indicated above, there is clear evidence that Na^+-K^+-ATPase is expressed in MRCs in the branchial epithelium of freshwater fishes or euryhaline (wide salinity tolerance) fishes in fresh water (teleosts and elasmobranchs). In addition, immunocytochemical studies of Na^+-K^+-ATPase (using heterologous antibodies) localised expression of the transporter to basolateral and/or tubulovesicular components of the MRCs. Partial and full-length Na^+-K^+-ATPase á-subunits have been cloned from a variety of freshwater species, and many of these studies have identified multiple isoforms. The basolateral expression of Na^+-K^+-ATPase is assumed to provide an exit step for the Na^+ from the branchial epithelial cell into the extracellular fluids. There is some evidence that one form of a Na^+-HCO_3^- cotransporter (NBC1) may be expressed in a subpopulation of MRC (defined by Na^+-K^+-ATPase immunolocalisation) in the acid-stressed Osorezan dace and the rainbow trout which could mediate both Na^+ and HCO_3^- exit to the blood. The basolateral exit step for Cl^- has not been studied directly, but an antibody raised against human CFTR localised to the basolateral regions of both MRC and PVCs in the freshwater killifish opercular epithelium, suggesting that this Cl^- channel may be involved.

Basolateral K^+ Recycling

The addition of 2 mM Ba^{2+} to the basolateral side of the killifish opercular epithelium inhibited the short-circuit current by 77%, suggesting the presence of K^+ channels to recycle the K^+ entering the cell via Na^+-K^+-ATPase. More recently, a full-length sequence for an inward rectifier K^+ channel (eKir) has been reported from Japanese eel gills, and immunogold staining with homologous antibodies has localised expression to the basolateral, tubular system in the MRC.

It is not clear if the uncertainty about the mode(s) of NaCl uptake is secondary to species differences, salinity differences, or the use of heterologous versus homologous antibodies, but more species need to be examined before a general model can be proposed. However, there is a distinct possibility that species differences do exist. In any case, the attending extrusion of H^+ (or NH_4^+) and HCO_3^- provides potential pathways for acid-base regulation and excretion of nitrogen. This is discussed further in section VI*D*.

Divalent Ion Uptake

Because freshwater concentrations of divalent ions such as Ca^{2+}, Mg^{2+}, and $Zn^{2+}$are far below plasma levels, fish must extract the needed ions from either food or freshwater itself. Fish (at least teleost) bone is acellular, so it does not provide a pool for Ca^{2+} as it does in terrestrial vertebrates. Recent evidence suggests that branchial uptake mechanisms exist for all three ions, but Ca^{2+} is the best studied.

Because the intracellular Ca^{2+} concentration of gill cells is presumably <1 μM, it is assumed that apical uptake is diffusional even from soft fresh water (10 μM Ca^{2+}), and partial or full-length sequences for a putative ECaC have now been reported from tilapia (B.-K. Liao, C.-H. Yang, T.-C. Pan, and P.-P. Hwang, unpublished data), fugu (GenBank accession) and rainbow trout. In both species, the putative ECaC is expressed in the gill, and in tilapia the expression increased fourfold when the fish were acclimated to low-Ca^{2+} fresh water, a condition where Ca^{2+} uptake increased (Liao et al., unpublished data). Presumed electrochemical gradients at the basolateral membrane suggest that active transport mechanisms must be involved for Ca^{2+}exit to the blood. Ca^{2+} transport across isolated membrane vesicles and the killifish opercular epithelium is Na^+ dependent, so Na^+/Ca^{2+} exchange may be involved. The complete sequence for a putative Na^+/Ca^{2+} exchanger (NCX) has been reported from tilapia (AY283779), but its expression in the gill of this species did not change under low Ca^{2+} conditions (T.-C. Pan, C.-H. Yang, C.-J. Huang, and P.-P. Hwang, unpublished data). Ca^{2+}-dependent ATP hydrolysis can also be measured in gill vesicle preparations, with a $K_{1/2}$ of 63 nM, so a high-affinity Ca^{2+}-activated ATPase is assumed to play a role. A partial sequence for a putative Ca^{2+}-activated ATPase (PMCA) has been reported (AF236669), but its expression in the gill did not change under low Ca^{2+} conditions (Liao et al., unpublished data). Taken together, these results suggest that apical uptake, rather than basolateral extrusion, is the limiting factor in Ca^{2+} balance, at least in tilapia. The fact that Ca^{2+}-activated ATPase activity covaries with Na^+-K^+-ATPase activity in the European eel and rainbow trout, both are stimulated by prolactin injection in the eel (224), and Ca^{2+} uptake is proportional to the density of MRC in tilapia, rainbow trout, and killifish suggest that MRCs are the site of Ca^{2+} uptake by these putative transport processes.

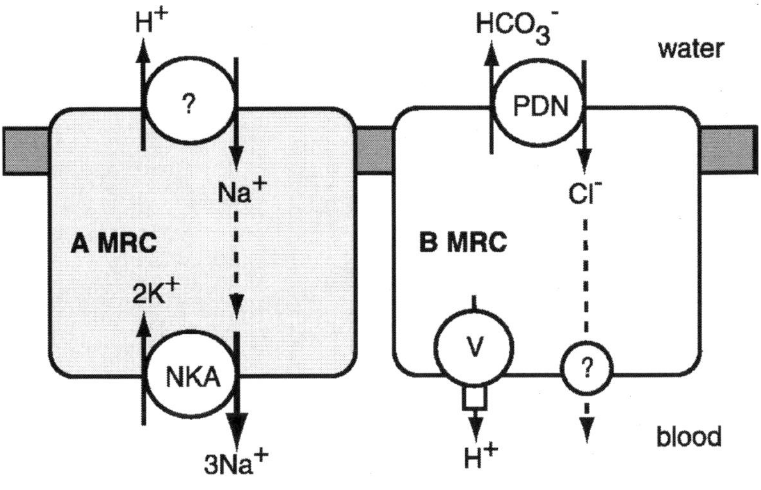

Figure: Working model for the Ca^{2+}uptake mechanisms across the freshwater teleost gill epithelium. Ca^{2+} is drawn in via a Ca^{2+} channel (ECaC) on the apical surface of the MRC and extruded across the basolateral membrane via a Ca^{2+}-ATPase (PMCA) or a Na^+/Ca^{2+}exchanger (NCE).

Water Permeability and Aquaporins

Early isotopic and physiological measurements of diffusional and osmotic water permeabilities suggested that both were higher in the branchial epithelium of freshwater than that of seawater fishes, and the substantial differential between the osmotic and diffusional permeabilities suggested the presence of "water-filled pores" in the freshwater, but not seawater, gill epithelium. Aquaporin-3 (AQP-3) now has been cloned from the gills of the eel and found to be 67–69% homologous to AQP-3 from rat, mouse, and human. The expression of AQP-3 declined significantly when eels were acclimated to seawater, and a homologous antibody localised AQP-3 immunoreactivity to both the apical and basolateral aspects of branchial MRC, as well as other branchial cells (PVCs not specified). AQP-3 also has been cloned from the gills of the acid-stressed, Osorezan dace, and homologous antibodies localised expression to the basolateral aspect of the MRC. What role AQP-3 plays in osmoregulation is unknown, although it may interact with CFTR. Moreover, another aquaporin (AQP-1) has been shown to transport both ammonia and CO_2, both of which traverse the fish branchial epithelium.

Lampreys

There is little known about the specifics of gill ion transport by lampreys in fresh water, but their plasma is hypotonic to fresh water, and more than one type of MRC is present in the gill epithelium of various lamprey species when acclimated to freshwater. A recent study has demonstrated two types of MRC in the Australian pouched lamprey: one that expresses Na^+-K^+-ATPase on the basolateral aspect and another that expresses CAII in the apical area and V-ATPase in the cytoplasm. The authors propose a model for parallel pathways for Na^+ and Cl^- uptake identical to that proposed for the elasmobranch gills and similar to that found in the proximal tubule cells (for Na^+ uptake) and the type B intercalated cells (for Cl^- uptake) of the mammalian kidney. These data are all consistent with the generally accepted hypothesis that these primitive agnathan fishes possess the same gill epithelial ionic transport mechanisms as those described for freshwater teleosts, suggesting that the origin of these pathways was of the order of 500 million years ago. For more general discussions of lamprey osmoregulation in fresh water.

Osmoregulation in Seawater

The majority of fish species live in the marine environment and thereby face the reverse osmotic and ionic gradients across the branchial epithelium to those found in the freshwater environment. Interestingly, the two major groups of modern, marine fishes (teleosts and elasmobranchs) have evolved different strategies to balance the potential osmotic loss of water and diffusional gain of NaCl. In teleosts, which are distinctly hyposmotic to seawater, the osmotic loss of water is balanced by ingestion of seawater and subsequent intestinal uptake of NaCl to withdraw needed water from the lumen contents. The resulting salt load adds to that produced by the diffusional uptake of NaCl across the gills.

The sum is balanced by gill epithelial NaCl extrusion, because the lack of a loop of Henle precludes the production of urine that is hyperosmotic to the plasma. In elasmobranchs, the retention of urea at extraordinary levels brings the plasma osmolarity hypertonic to seawater, so oral ingestion of seawater is not necessary. Nevertheless, sharks, skates, and rays must offset the diffusional uptake of NaCl across the gill epithelium by excretion of a hypersaline fluid by the rectal gland, derived from the posterior intestine. However, various studies have shown that the rectal gland is not a vital organ because fish were able to osmoregulate after removal of the rectal gland, despite producing urine that is isotonic to the plasma. Hagfish are isotonic to seawater, and the ionic differentials (especially for divalent ions) are probably maintained by renal and mucous secretion. Lampreys, on the other hand, maintain substantial ionic gradients across the gill epithelium, and it appears that their osmoregulatory strategies are similar to those found in marine teleosts. For general reviews of marine fishosmoregulation, for specific reviews on mechanisms of gill NaCl transport.

NaCl Secretion in Marine Teleosts

Net salt extrusion by the fish gill epithelium was first described over 70 years ago in a classic paper on Cl^- transport across the perfused gills of the eel by Ancel Keys, in which he stated: "These experiments demonstrated beyond doubt that a concentrated chloride solution is secreted by the gills in opposition to a large concentration gradient." In a subsequent paper, Keys stated that "chloride secretion exhibited by the gills of the eel in seawater is an active process requiring the expenditure of large amounts of energy," surely one of the earliest suggestions that an epithelium could actively transportions. This latter paper is also a classic in the field, because it was the first description of a specific cell involved in active transport: the "Cl^- secreting cell," which is now known to be a type of teleost MRC.

The mechanisms of NaCl secretion by at least the teleost gill epithelium are more clear than the mechanisms of NaCl uptake by the same tissue. As the importance of Na^+-K^+-ATPase in a variety of cells and epithelia (including thefish gills) became known 35 years ago, it was proposed that apical Na^+-K^+-ATPase-mediated Na^+ extrusion across the branchial epithelium, because, for example, removal of external K^+ or external application of ouabain inhibited Na^+ efflux, measured radioisotopically.

This proposition of apical Na^+-K^+-ATPase extruding cellular Na^+ was rendered untenable when Karnaky demonstrated unequivocally that Na^+-K^+-ATPase was restricted to the basolateral membrane of the MRC in the branchial epithelium of the killifish, and this localisation has now been confirmed for a variety of species, including elasmobranchs. The importance of basolateral Na^+-K^+-ATPase in both Na^+ and Cl^- extrusion was demonstrated by the fact that injection of ouabain into the eel completely inhibited both Na^+-K^+-ATPase activity and the efflux of both ions.

It was proposed that NaCl extrusion was mediated by a basolateral cotransport of NaCl down the electrochemical gradient produced by Na^+-K^+-ATPase, coupled with an apical extrusion of Cl^- via a channel and paracellular extrusion of Na^+, both down their respective electrochemical gradients. This model was consistent with work on other epithelia and was corroborated by concurrent, electrophysiological studies of the epithelial sheets from the operculum of the killifish and the lower jaw of the longjaw mudsucker.

The killifish opercular epithelium contained ~60% MRCs, maintained an open-circuit electrical potential of 19 mV (serosal side positive) and an electrical resistance of 174 Ù · cm², and produced a net short-circuit current (I_{sc}; 137 μA/cm²) that was equivalent to the net Cl^- efflux, measured isotopically. Addition of ouabain or furosemide, or ionic substitutions to the basolateral side, inhibited the I_{sc} and effluxes of Na^+ and Cl^-, consistent with the model of basolateral Na^+-K^+-ATPase and NaCl cotransport. Studies showing a direct correlation between MRC number and Cl^- currents in the opercular and jaw skin epithelia, combined with salinity-induced changes in structure and number of MRCs in fish gills suggested the primacy of the MRC in NaCl extrusion. This was proven directly when Cl^- currents were localised to the MRC of the killifish operculum, using a vibrating probe technique.Below diagrams the generally accepted model for

NaCl extrusion by the teleost MRC, and supporting evidence for specific pathways follows.

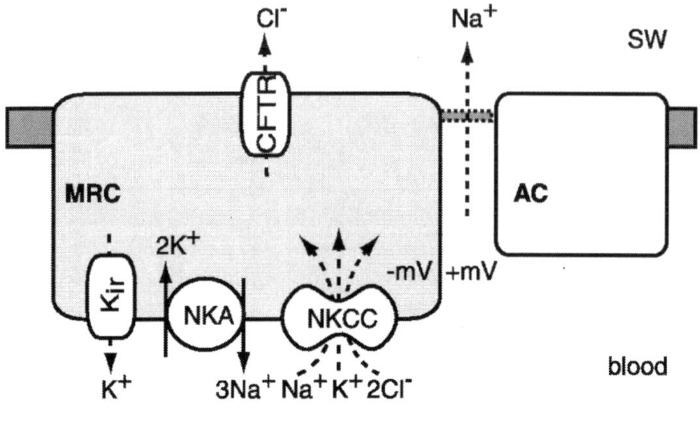

Figure: Working model for the extrusion of NaCl by the marine teleost gill epithelium. Plasma Na^+, K^+, and Cl^- enter the cell via basolateral NKCC; Na^+ is recycled back to the plasma via Na^+-K^+-ATPase and K^+ via a K^+ channel (K_{ir}). Cl^- is extruded across the apical membrane via a Cl^- channel (CFTR). The transepithelial electrical potential across the gill epithelium (plasma positive to seawater) drives Na^+across the leaky tight junctions between the MRC and the AC.

Basolateral Na^+ and Cl^- Uptake

NA^+-K^+-ATPASE

Complete or partial sequences for Na $^+$-K^+-ATPase now have been reported from a variety of freshwater and euryhaline teleosts as well as stenohaline marine species from the Antarctic and New Zealand, where multiple isoforms were identified. Na^+-K^+-ATPase expression can be localised to the MRCs in the gill epithelium in marine species (including the dogfish shark) or euryhaline species in seawater (including the stingray) using nucleotide probes and antibodies. In some cases, Na^+-K^+-ATPase has been specifically localised to the basolateral aspect of the MRC Expression and activity of Na^+-K^+-ATPase is often correlated directly with salinity oreover, relative Na^+-K^+-ATPase activity in gill biopsies can actually predict the future downstream migratory behaviour in freshwater brown trout. On the other hand, Na^+-K^+-ATPase activity and/or expression actually decreases when some euryhaline species are acclimated to seawater. The expression of Na^+-K^+-ATPase increased (compared with seawater controls) when the sea bass (*Dicentrarchus labrax*) is acclimated to either fresh water or 200% seawater, which suggests that any osmotic stress might affect the expression of this important transport enzyme. Some of these discrepancies may be due to differential expression of Na^+-K^+-ATPase isoforms, because a recent study has demonstrated that, in the rainbow trout gills, Na^+-K^+-ATPase a1b is upregulated in seawater, but Na^+-K^+-ATPase á1a is downregulated.

NA-K-2CL Cotransport

Fragments of the NKCC gene have been cloned from the European eel (homologous to NKCC1 and NKCC2), the brown trout and Atlantic salmon (NKCC1), and the striped bass (*Morone saxatilis*). In all cases, transcripts were detected in gill tissue (by Northern blots), and expression was highest when the fishes were in seawater. The T_4 antibody raised against human colonic NKCC1 has localised expression of the cotransporter to the basolateral aspect of MRC in the gill epithelium of the giant mudskipper (*Periophthalmodon schlosseri*) and Hawaiian goby (*Stenogobius hawaiiensis*), in MRC in the Atlantic salmon (colocalised with Na^+-K^+-ATPase) and killifish and in what appear to be MRC in the brown trout. In the latter three studies, the protein expression was directly correlated with salinity, as it was (measured by Western blots) in gill extracts from striped bass in seawater versus fresh water.

Basolateral K^+ Recycling

The hypothesis that K^+ that enters the cell associated with Na^+-K^+-ATPase and NKCC is recycled through the K^+ channel is suggested by the effect of Ba^{2+} on the I_{sc} across the killifish operculum and the cloning of an inward rectifier K^+ channel in the Japanese eel. Acclimation of the eel to seawater increased the expression of the gene product significantly, as measured by RNase protection assay.

Apical Salt Extrusion

Marshall's group has identified a low-conductance (8 pS) anion channel in a primary culture of cells from the killifish opercular epithelium (which retained MRC) that was stimulated by cAMP and inhibited by two anion channel blockers, diphenyl-amine-2-carboxylic acid (DPC) and 5-nitro-2-(3-phenylpropylamino)benzoic acid (NPPB). A full-length clone of a CFTR gene was sequenced from a killifish gill cDNA library that encoded a protein (59% identical to human CFTR amino acid sequence) and that produced a cAMP-activated anion conductance when its mRNA was expressed in *Xenopus* oocytes. Importantly, expression of the product in the gills increased severalfold after transfer of killifish to seawater.

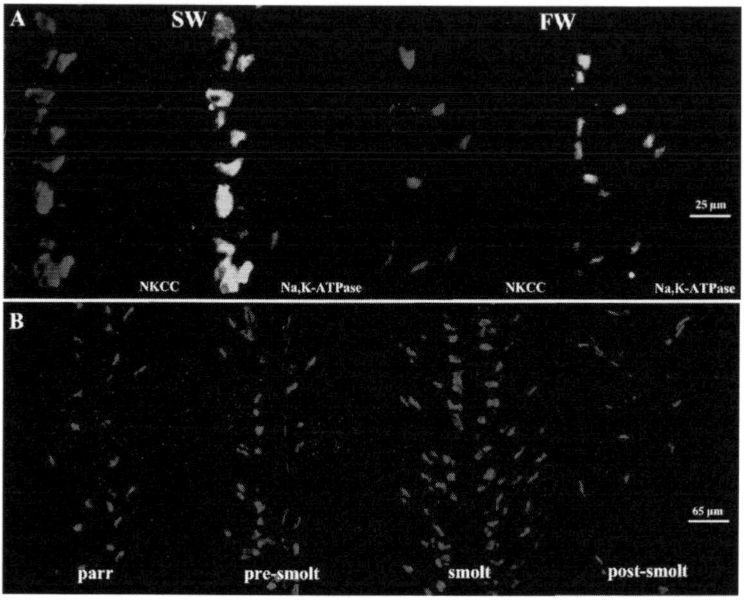

Figure: Confocal laser scanning images of kfCFTR immunofluorescence of mouse anti-human CFTR and goat anti-mouse IgG Oregon Green 488 in killifishopercular epithelium show the distribution of kfCFTR (green fluorescence + red yields yellow) in mitochondria-rich (MR) chloride cells counterstained for mitochondria with Mitotracker Red. *A*: membrane from a SW-adapted animal has kfCFTR immunofluorescence at 4.5 µm from the surface of the chloride cells is localised to the apical crypt (white arrows). The asterisk indicates a gap in the pavement cell layer (*B*). The same frame as for *A* but at a depth of 7.5 µm at the plane of the nucleus of the MR cell. Blue arrows in *B* indicate the mitochondria-rich chloride cells below the apical crypt. *D* and *E*: similar kfCFTR immunostaining, but in MR cells from fish acclimated to FW. An optical section (*D*) at 4.5 µm from the surface has positive kfCFTR immunofluorescence evenly distributed in the chloride cells (white arrows) as well as in the pavement cells at the plane of the pavement cell nuclei (orange arrow). Optical section (*E*) is same frame as for *D* but 10.5 µm into the tissue at the level of the MR cell nuclei. Blue arrows indicate the MR cells with diffusely distributed positive immunofluorescence for kfCFTR. *F* and *G*: similar immunostaining but for a fish transferred from FW to SW for 48 h. *F*: an optical section 9.0 µm from the surface with ring-shaped kfCFTR positive immunofluorescence (white arrows). *G*: 13.5 µm from the surface with kfCFTR immunofluorescence diffusely distributed in MR cells (blue arrows). Scale bars in *A*, *D*, and *F* are 20 µm.

Full-length clones of CFTR are now available from the Atlantic salmon (95% identical to killifish amino acid sequence) and fugu (*Fugu rubripes*, 84% identical to killifish amino acid sequence).

Antibodies directed against hCFTR have localised expression to the apical region of MRC in the giant mudskipper and Hawaiian goby, and a monoclonal antibody directed against the shared carboxy terminus of hCFTR and kCFTR confirmed this location in the killifish gill epithelium. In the latter study, the immunoreactive CFTR moved from the central part of the cytosol to the apical surface of the MRC when the killifish were acclimated to seawater.

In the Hawaiian goby, CFTR-positive cells increased fivefold in seawater-acclimated fish. For an excellent review of the role of CFTR in the seawater teleost gill.

Does the Elasmobranch Gill Excrete NaCl?

The fact that removal of the rectal gland from at least *Squalus acanthias* does not result in significant hypernatremia suggests that other tissues can secrete salt when necessary.

The elasmobranch gill epithelium does contain MRCs, which express immunoreactive Na^+-K^+-ATPase, and NKCC expression has been noted in the dogfish gills.

However, neither MRC number or size nor Na^+-K^+-ATPase activity increased after the removal of the rectal gland in the dogfish, counter to what one would expect if branchial extrusion mechanisms were upregulated in response to the loss of the rectal gland.

Moreover, acclimation of freshwater Atlantic stingrays to seawater was associated with a decrease in branchial Na^+-K^+-ATPase activity and Na^+-K^+-ATPase immunoreactivity. More studies are warranted, but it is possible that increased renal Na^+ and Cl^- excretion could offset diffusional salt uptake in the absence of the rectal gland, because the elasmobranch gains water by osmosis and, therefore, has high urine flow rates.

As long as the rate of water absorption via osmosis is greater than or equal to diffusional salt uptake, blood osmolarity can be maintained by excretion of urine that is isotonic or even hypotonic to the plasma. Indeed, Burger found increased urinary salt loss after removal of the rectal gland in the spiny dogfish.

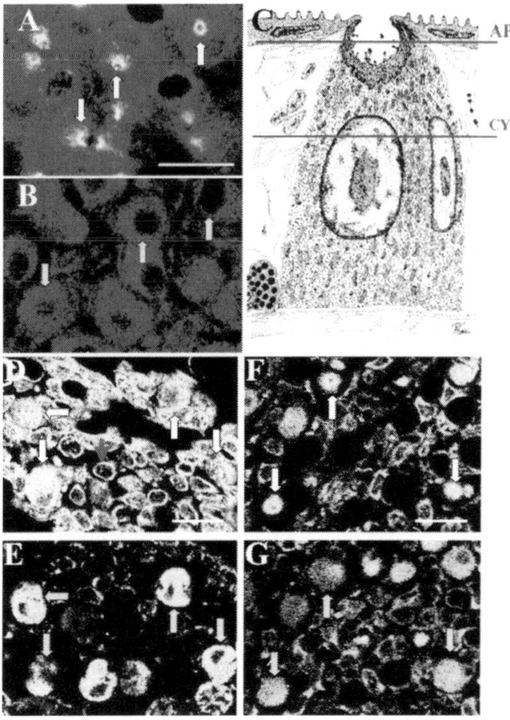

Figure: Confocal laser scanning images of kfCFTR immunofluorescence of mouse anti-human CFTR and goat anti-mouse IgG Oregon Green 488 in killifishopercular epithelium show the distribution of kfCFTR (green fluorescence + red yields yellow) in mitochondria-rich (MR) chloride cells counterstained for mitochondria with Mitotracker Red. *A*: membrane from a SW-adapted animal has kfCFTR immunofluorescence at 4.5 μm from the surface of the chloride cells is localized to the apical crypt˙ (white arrows). The asterisk indicates a gap in the pavement cell layer (*B*). The same frame as for *A* but at a depth of 7.5 μm at the plane of the nucleus of the MR cell (see drawing, *C*, for approximate depths of the optical sections; AP, apical crypt level; CY, cytosol level). Blue arrows in *B* indicate the mitochondria-rich chloride cells below the apical crypt. *D* and *E*: similar kfCFTR immunostaining, but in MR cells from fish acclimated to FW. An optical section (*D*) at 4.5 μm from the surface has positive kfCFTR immunofluorescence evenly distributed in the chloride cells (white arrows) as well as in the pavement cells at the plane of the pavement cell nuclei (orange arrow). Optical section (*E*) is same frame as for *D* but 10.5 μm into the tissue at the level of the MR cell nuclei. Blue arrows indicate the MR cells with diffusely distributed positive immunofluorescence for kfCFTR. *F* and *G*: similar immunostaining but for a fish transferred from FW to SW for 48 h. *F*: an optical section 9.0 μm from the surface with ring-shaped kfCFTR positive immunofluorescence (white arrows). *G*: 13.5 μm from the surface with kfCFTR immunofluorescence diffusely distributed in MR cells (blue arrows). Scale bars in *A*, *D*, and *F* are 20 μm.

Divalent Ion Excretion

Although the Ca^{2+} and Mg^{2+} concentrations of the plasma of marine fishes are below that of seawater, it is unlikely that diffusional influx is significant because of the serosal positive electrical potential across the gills. However, the oral ingestion of seawater by at least marine teleosts presents a divalent ion load to the fish (to the degree that intestinal uptake exists), which must be excreted by renal and branchial mechanisms.

For instance, Hickman calculated that only ~15% of the ingested Mg^{2+} (and SO_4^{2-}) was absorbed, but 100% of that was excreted renally by the southern flounder (*Paralichthys lethostigma*). In contrast, 68% of the ingested Ca^{2+} was absorbed, and only 11% of that could be accounted for by renal loss. Thus the remaining 89% of the Ca^{2+} removed from the intestine must be excreted by the gills. This putative branchial extrusion of Ca^{2+} could be passive because the calculated Nernst equilibrium electrical potential for Ca^{2+} is of the order of 10–15 mV (inside positive), which is probably below that across the marine fish gill epithelium. More data are needed.

Hagfish and Lampreys

Despite the lack of significant osmotic or NaCl gradients across their gills, the hagfish branchial epithelium does contain cells that share many of the morphological characteristics that define the MRC of teleosts and elasmobranchs. Moreover, Na^+-K^+-ATPase can be immunolocalised to MRC on the filaments and lamellae of the gill epithelium of the Atlantic hagfish, and NHE-1 expression was detected in gill extracts by Western blots in the same species. Intracellular distribution was not determined, but expression of these two proteins suggests that the MRC in the gills of the hagfish may be involved in acid-base regulation, as it is in teleosts and elasmobranchs.

All lampreys breed in fresh water, and one group, the petromyzonitids, enter seawater after metamorphosis. The marine lampreys (e.g., *P. marinus*) migrate back to fresh water as breeding adults, and one population is actually resident in the Great Lakes of North America. Little is known about osmoregulation in the marine populations because of the difficulty of capturing animals, but MRCs in the gill epithelium of marine lampreys lose basolateral invaginations after acclimation to fresh water, and the activity of gill Na^+-K^+-ATPase is much higher in seawater versus freshwater individuals, suggesting a role for Na^+-

K^+-ATPase in MRCs in seawater osmoregulation. More data are needed, but it is generally assumed that the gill mechanisms for osmoregulation in seawater lampreys parallel those in the marine teleost gills. Modern molecular techniques should allow confirmation of this assumption.

Euryhalinity

The vast majority of fishes only can tolerate salinities near to either fresh water or seawater (termed stenohaline), but a large number of species (in all major fishgroups) are able to tolerate a much broader range of salinities, including some that are fully euryhaline (e.g., lampreys, Atlantic stingrays, bull sharks, sturgeon, eels, salmon, shad, striped bass, mullet, flounders, killifish, etc.).

Recent evidence suggests that the specific molecular or biochemical events that accompany acute salinity transfers may be species specific, but they generally involve reciprocal regulation of expression and/or activity of Na^+-K^+-ATPase, NKCC, CFTR, V-ATPase, and possibly NHE. One might propose that stenohaline species lack the requisite, branchial salt transport mechanisms and/or are unable to make the necessary quantitative changes in renal or intestinal function for osmoregulation in seawater versus fresh water.

However, there is now clear evidence that the putative branchial salt absorption mechanisms (NHEs, and/or V-ATPase electrically balanced by an ENaC-like channel) are present in marine species, presumably for acid-base regulation.

This was originally suggested as an evolutionary "holdover" from the ancestors of present-day marine gnathostomes that inhabited fresh water, but this idea was challenged when Na^+ and Cl^- substitutions were shown to also inhibit acid and base excretion, respectively, from the Atlantic hagfish, an osmoconforming agnathan that is physiologically limited to seawater. Hagfish are believed to lack a period of freshwater existence in their evolutionary history, suggesting that Na^+- and Cl^--linked acid and base secretory mechanisms originally evolved for acid-base regulation in seawater, and not for ion absorption in fresh water.

If the ionic transport mechanisms that mediate ion absorption in modern freshwater fishes arose in the earliest vertebrates for acid-base regulation and persist today in modern marine fishes (including relatively stenohaline marine teleosts, elasmobranchs, and hagfish), one might ask why all fishes are not euryhaline? Presumably, an inability to alter the secretion of hormones and paracrines that control the branchial processes of osmoregulation (in addition to intestinal and renal processes) may be a factor, but it seems likely that the interplay between branchial ionic permeability and the efficacy of ionic transport may be a major limiting factor in at least the transition between seawater and brackish or fresh water. The importance of permeability is best seen in the role of Ca^{2+} in freshwater survival of various species of inshorefishes that are normally stenohaline, as well as in the Na^+ balance and survival of the marine pinfish (*Lagodon rhomboides*) in low salinities.

The inability of most freshwater fish species to enter hyperosmotic salinities simply may be caused by the lack of the requisite extrusion proteins in their MRC, or more subtle hormonal, renal, or intestinal deficiencies may be present (e.g., no neuroendocrine axis to stimulate oral ingestion). Nevertheless, there is a substantial number of fish families that have euryhaline members, including such relatively primitive fish species as lampreys, sturgeons, eels, tarpon, and rainbow trout, suggesting that euryhalinity evolved several times in the early vertebrate lineage.

Osmoreceptors

The ability of some euryhaline species to enter lower salinities during seasonal migration (e.g., eels, salmon) or daily movements in the inshore environment (e.g., killifish) suggests that relatively rapid changes in transport can take place, and that osmo- or ionoreceptors may be present in the branchial epithelium, vasculature, or central nervous system.

There is now good evidence that the gill epithelium itself can respond to changes in the external salinity and/or composition. It was found that a 50 mosM increase in the Ringer solution bathing the basolateral (but not apical) side of the killifish opercular epithelium produced a significant decline in the volume of individual MRC, which was associated with a doubling of the I_{sc}.

The increased I_{sc} could be inhibited by the addition of bumetanide, suggesting that activation of NKCC was involved in the response to basolateral hypertonicity. More recently, another study demonstrated that a 40 mosM decrease in the basolateral solution inhibited the I_{sc} by 58%, which corroborates the hypothesis that plasma osmolarity changes can elicit very rapid changes in at least active NaCl extrusion by the killifish gill epithelium.

The intracellular messengers that elicit these responses are relatively unstudied, but there is now evidence that protein kinase C (PKC) and myosin light-chain kinase (MLCK), but not protein kinase A (PKA), are activated by increased osmolarity, and that a "heavily serine phosphorylated protein of about 190 kDa" is also in higher concentrations is seawater- but not fresh water-acclimated killifish, suggesting that upregulation of a serine/threonine kinase may be involved.

On the other hand, the activity of three mitogen-activated protein kinases (SAPK2, SAPK1, and ERK1) all increased significantly in the gill epithelium during hyposmotic stress of the killifish. Kültz et al. have cloned a 14.3.3 protein from the killifish that is expressed in high concentrations in the gill epithelium, and which increases two- to fourfold within 24 h of transfer of the fish into seawater.

This suggests a role for this putative cofactor for protein kinases, phosphatases, and other phosphoproteins in structural or functional changes in the epithelium during at least short-term salinity changes. Longer term morphological and physiological changes associated with euryhalinity are discussed in section VIII.

Recent studies suggest that polyvalent cation receptors (CaR) in the gill epithelium also may stimulate osmoregulatory changes by responding to changes in internal or external [Ca^{2+}]. Full-length cDNA sequences of CaR have now been reported from both the gilthead sea bream (*Sparus aurata*) and spiny dogfish, and nucleotide probes have localised expression in branchial MRCs in both species, as well as the winter flounder (*Pleuronectes americanus*) and Atlantic salmon. Moreover, the CaR also may be a salinity receptor, because (after expression in human embryonic kidney cells) its sensitivity to Ca^{2+} or Mg^{2+} was inversely related to the Na^+ concentration of the incubation medium. Thus the MRC itself may be a salinity sensor, responding to changes in volume and/or changes in stimulation of Ca^{2+} receptors.

pH Regulation

Fishes, like all vertebrates, have three compensatory mechanisms to regulate the acid-base status of their extracellular body fluids: *1*) instantaneous physiochemical buffering with bicarbonate and nonbicarbonate buffers, *2*) respiratory adjustments of the "open" CO_2-HCO_3^- buffer system within minutes, and *3*) net transport of acid-base relevant molecules between the animal and environment within minutes to hours. Using an aqueous medium for respiration limits the first two mechanisms relative to air-breathing vertebrates, but enhances the third mechanism.

Respiratory Compensation

In well-aerated water, the obligatorily high gill ventilation rates and high capacitance of CO_2 allows metabolically generated CO_2 to readily leave through the gill without a substantial buildup in the tissues of fishes, as it may in air-breathing vertebrates. This results in an arterial PCO_2 that is never more than a few mmHg above the water irrigating the gills (Pa_{CO2} ~1 mmHg in normocapnic water), a minute CO_2 gradient compared with air-breathing vertebrates such as mammals. Accordingly, the blood bicarbonate concentration of fishes is also low (~4 mM in normocapnic water compared with 24 mM in humans)The low steady-state Pa_{CO2} and bicarbonate concentration minimise the ability of fishes to "blow-off" buffered metabolic acid loads via hyperventilation as occurs in mammals and reduces the buffering capacity of the extracellular compartment during metabolic acid-base disturbances.

For example, data on longhorn sculpin, lemon sole (*Parophrys vetulus*), and rainbow trout show that acute mineral acid infusions actually cause a slight respiratory acidosis (in addition to a severe metabolic acidosis), suggesting a lack of any respiratory compensation. The same was shown for several species when lacticacidosis was induced by strenuous exercise; a slight respiratory acidosis (Pa_{CO2} increase of 2–3 mmHg) developed that lasted for 3–10 h even though hyperventilation occurred in response to internal hypoxia.

Although respiratory compensation for metabolic acidosis does not appear to exist in fishes, several elasmobranch and teleost species have been shown to hyperventilate in response to environmental hypercapnia.

However, this mechanism is not an effective compensation for respiratory acidosis for three reasons: *1*) it only responds to hypercapnia of external origin, and not to hypercapnia of endogenous origin; *2*) even if hyperventilation was pronounced enough to abolish the Pv_{CO2} gradient between blood and environmental water completely, the blood pH would only be raised by one- or two-tenths of a pH unit (calculated from Henderson-Hasselbalch equations) above what it would be in the absence of hyperventilation because of the small steady-state PCO_2 gradient at the gills; and *3*) hyperventilation can never fully compensate for an environmental hypercapnia, because the source of the acidosis is the respiratory medium. Therefore, the PCO_2-sensitive hyperventilation is more likely to have evolved as a "back-up" of PO_2-sensitive hyperventilation to maintain O_2 uptake at the gills, because hypoxia and hypercapnia usually occur simultaneously in natural waters.

Metabolic Compensation

The limitations that water-breathing imposes on bicarbonate buffering and respiratory compensations make fishes heavily dependent on net exchange of nonvolatile acid-base equivalents with their environment, a task well suited to gills. As noted earlier, fish gills are covered with an epithelium that is in direct contact with an aqueous external medium that is usually large enough to act as an infinite source and sink of solutes (i.e., acid-base equivalents and ions with which they are exchanged).

Assuming an appropriate ionic environment (e.g., available counter ions and near neutral pH), this is an advantage over internal transport tissues such as the kidneys, which could build-up opposing gradients because of the limited volume of the urine. For example, resting water flow rates at the gills are between 5,000 and 20,000 ml $\cdot$ kg^{-1} $\cdot$ h^{-1} which dwarfs the urine flow rates of fishes (~1–10 ml $\cdot$ kg^{-1} $\cdot$ h^{-1}) and other vertebrates (e.g., humans, ~1 ml $\cdot$ kg^{-1} $\cdot$ h^{-1};) by several orders of magnitude.

These high flow rates presumably minimise the creation of large pH gradients across the gills during times of branchial net-acid excretion and usually obviate the need for complementary secretion of buffers such as phosphate and ammonia, which are required by kidneys. Accordingly, although the skin, rectal glands, and kidneys of fishes have all been implicated in net acid-base transport, the gills usually account for over 90% of compensatory net acid-base transfers in fishes.

Data indicate that most fishes secrete a nonvolatile, net acid flux of 10–100 μmol $\cdot kg^{-1}$ $\cdot h^{-1}$ under control steady-state acid-base conditions, with a few exceptions where a net base flux of similar magnitude is observed. This "typical" control net acid flux is comparable to that of mammals such as humans (~50 μmol $\cdot$ kg^{-1} $\cdot$ h^{-1}), and presumably reflects a net production of nonvolatile acidic equivalents from several metabolic processes. In nature, fishes experience large deviations from these control values caused by either external stressors, such as changes in respiratory gas partial pressures (e.g., hypercapnia, hypoxia, and hyperoxia), pH, salinity, and temperature, or by internal stressors such as lacticacidosis (e.g., during prey capture, predator avoidance, and migrations). By fitting fish with indwelling blood catheters, and placing them in recycling water containers of fixed volume, the internal acid-base responses and whole animal compensatory net acid fluxes to these stressors have been studied extensively in teleosts and elasmobranchs, and to a lesser degree in lampreys and hagfish. For details on the blood acid-base status, and time courses of compensatory transport.

In general, regardless of the fish taxon or the source of the acid-base stress, the return of blood toward control pH is primarily due to adjustments of blood bicarbonate concentrations via exchange of acid-base equivalents at the gills.

Although variable with the type and extent of acid-base disturbance, compensatory transport is usually activated within 20–30 min. of the disturbance and can reach net-acid excretion rates of 1,000 $\mu mol \cdot kg^{-1} \cdot h^{-1}$ within a few hours of inducing acidosis and net-base excretion rates as high as 1,000 $\mu mol \cdot kg^{-1} \cdot h^{-1}$ within a few hours of inducing alkalosis Therefore, fish gills respond to acid-base disturbances rapidly and efficiently, demonstrating well-developed mechanisms of acid-base transport that are under tight regulation.

The mechanisms of ion and acid-base transport across the gill epithelium have been the subject of many studies in the past 70 years, beginning with the first studies by Homer Smith and August Krogh. Early in vivo experiments correlated isotopic fluxes of Na^+ and Cl^- to fluxes of net acid and ammonia across the gills of freshwater teleosts and demonstrated that secretion of endogenous acid (H^+ and/or NH_4^+) is linked to transepithelial absorption of Na^+ and that secretion of endogenous base (HCO_3^- and/or OH^-) is independent of acid secretion and linked to transepithelial absorption of Cl^- Although the goal of these early studies was to determine the mechanism of ion absorption in freshwater fishes, the model also provided mechanisms for secreting ammonia and regulating the metabolic component of extracellular acid-base status (i.e., strongion difference). It was originally assumed that seawater fishes would not possess these transport mechanisms, because of the obvious osmotic and ionic burden that active Na^+ and

Cl^-absorption would impose on the active NaCl secretion mechanisms. However, in vivo studies on seawater fishes demonstrated similar links between Na^+ and Cl^- absorption and acid and base secretion. These include a demonstration of complete inhibition of net-acid excretion from two marine elasmobranchs [spiny dogfish and little skates (*Raja erinacea*)] and two marine teleosts (toadfish and longhorn sculpin) when external Na^+ was substituted with the organic ions choline or *N*-methyl-D-glucamine, and an increased net-acid excretion from longhorn sculpin (presumably due to decreased HCO_3^- secretion) when external Cl^- was substituted by isothyanate.

The presence of external Na^+- and Cl^--linked acid and base secretory mechanisms in seawater fishes is now generally accepted, despite the incurred ionic load. Regardless of the mechanisms, these Na^+- and Cl^--linked acid-base transport mechanisms make acid-base regulation sensitive to external salinity.

For example, seawater teleosts generally compensate for acidosis faster than freshwater teleosts when presented with similar pH disturbances. Recently, we examined the effect of salinity on compensation from hypercapnia in the euryhaline Atlantic stingray and demonstrated that salinity has the same effect on rate of recovery in elasmobranchs. In two of these studies, $[HCO_3^-]$ in the water was held constant, ensuring that the effect of salinity was due to the availability of Na^+ and Cl^- counterions, and not bicarbonate in the water at the surface of the gills.

The universal finding that acid secretion is linked to Na^+ absorption, and that base secretion is linked to Cl^- absorption, does not mean that all fishes possess the same arrangement of acid-base transport proteins in their gills. In fact, its now clear that the type of acid-base transport proteins expressed, and their locations (cellular and subcellular), vary with species and with external water conditions, such as salinity and pH. Many studies have focused on the identification and location of proteins involved in acid secretion (e.g., V-ATPase, Na^+-K^+-ATPase, NHEs, anion exchangers, CA, NBC) with pharmacological, immunological, and molecular techniques.

These studies support the presence of two acid secretion mechanisms in fishes, which are also the two proposed mechanisms of Na^+ absorption: *1*) an apical V-ATPase electrically linked to Na^+ absorption via ENaC-like channels, and *2*) electroneutral exchange of Na^+ and $H^+(NH_4^+)$ via proteins of the NHE family. Fewer studies have focused on the identification and location of proteins involved in base secretion, but they suggest the presence of two apical Cl^-/HCO_3^- exchangers in fishes: *1*) AE1 and *2*) pendrin.

Acid Secretion

V-ATPase

Pharmacological agents have been used to determine the possible mechanisms of acid secretion by the gills of several fish species. In addition to blocking Na^+absorption, amiloride has been shown to inhibit net-acid excretion when placed in water irrigating the gills of rainbow trout, brown trout, European flounder, killifish, little skate, and longhorn sculpin. However, these results do not distinguish between the two acid secretion mechanisms, because amiloride inhibits Na^+ transport by NHEs and ENaC.

For example, amiloride would be expected to inhibit acid secretion from an NHE by blocking the exchanger directly, but could also theoretically inhibit acid secretion by inhibiting an ENaC-like channel and thereby altering the apical electrochemical gradient. However, three studies were able to eliminate NHE as the predominant acid transporter in three freshwater teleost species.

In rainbow trout, 0.1 mM amiloride had no effect on net acid excretion, a concentration that inhibits >80% of Na^+ absorption in this species. Similarly, 1 mM amiloride had no effect on net acid excretion from brown trout but did inhibit ~70% of Na^+ absorption. Finally, 5 mM 5-(*N*,*N*-dimethyl)amiloride (HMA), a specific inhibitor of NHEs, had no effect on net acid excretion (or Na^+ absorption) from isolated gill filaments of freshwater European flounder. These demonstrations of decoupled acid secretion and Na^+ absorption are strong arguments against a direct role for apical NHEs in freshwater teleosts, and together with the apical localisation of V-ATPase in rainbow trout, coho salmon, and tilapia, help support the role of V-ATPase as the predominant mechanism of acid secretion in freshwater teleosts.

Several studies have attempted to determine if the quantity of V-ATPase expression in freshwater teleost gills is increased during acidosis, which would be expected if this proton pump were responsible for the majority of acid secretion. In support of this hypothesis, a recent study used real-time PCR (RT-PCR) on rainbow trout gills to demonstrate large, sustained increases (4- to 70-fold increases) in V-

ATPase expression during intravascular infusions of HCl. Unfortunately, the results for hypercapnia are conflicting, even though almost all of them were conducted on a single family of fishes (Salmonidae, mainly rainbow trout). For example, Northern blots and RT-PCR detected a transient 1.5- to 2.0-fold increase in B subunit expression in the first 2 h of chronic hypercapnia, but this increase subsided to control levels by 4 h and was never again elevated for the rest of the experiment (24 h total).

Other studies, using the same species and subunit, suggested increased mRNA and protein expression after 18 h of hypercapnia, a time frame that is out of phase with the RT-PCR and Northern blot results. In contrast, a preliminary study on rainbow trout failed to detect any change in A subunit protein levels after 40 h of hypercapnia (405), and a study on another salmonid (Atlantic salmon) actually measured a small, but consistent, decrease in B subunit mRNA expression after 24 h of 2% hypercapnia.

Unfortunately, the reasons for these inconsistent results are unknown, but they do suggest that large changes in the expression of V-ATPase probably are not an important mechanism of regulating net acid secretion during hypercapnia. Interestingly, analogous tissues (such as mammalian renal tubule collecting ducts) appear to use rapid changes in the location (vesicular vs. apical) and not the total expression of V-ATPase pumps, as mechanisms of regulating acid secretion.

Prior to studies that localised V-ATPase and measured changes in the enzyme's expression, a series of ultrastructural studies were conducted to determine if morphological changes to PVC and/or MRC occur in response to acid-base disturbances, as they do in mammalian renal tubule intercalated cells.

Scanning electron micrographs demonstrated a 30 and 85% decrease in apical MRC fractional area (MRCFA) during hypercapnia in rainbow trout and brown bullhead catfish, respectively. They were also used to measure a 50% increase in MRCFA during metabolic alkalosis in rainbow trout. These changes in MRCFA were directly proportional to changes in unidirectional Cl^- uptake and inversely proportional to changes in net acid secretion and unidirectional Na^+ uptake, suggesting that MRCs are responsible for apical Cl^-/HCO_3^- exchanges and not for acid secretion or Na^+absorption. Alternatively, a 100% increase in the density of PVC microvilli was measured in response to hypercapnia in brown bullhead catfish, suggestive of an increase in apical cell area of this cell type. It was proposed that net acid secretion was increased by removal of Cl^-/HCO_3^- exchangers from the apical membrane of MRC and/or insertion of proton pumps into the apical membrane of PVCs, a model that is consistent with the expression of apical Cl^-/HCO_3^- exchangers in MRCs and V-ATPase in PVCs.

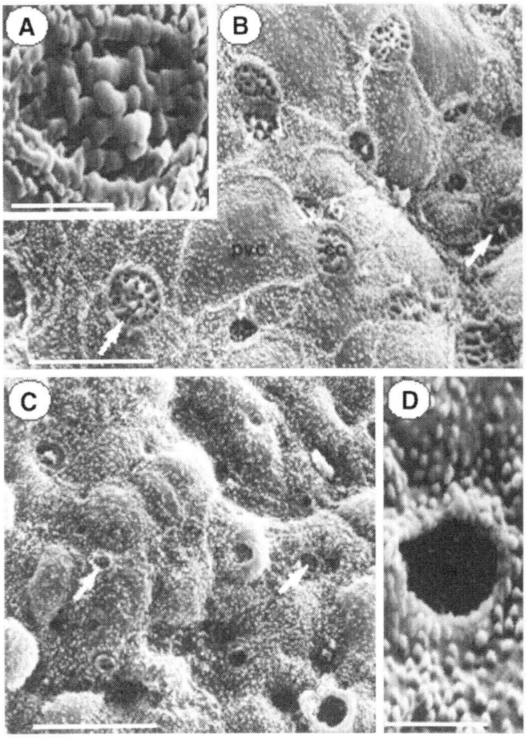

Figure: Representative low- and high-magnification scanning electron micrographs of the filamental epithelium from brown bullhead catfish (*Ictalurus nebulosus*) during normocapnia (*A* and *B*) and after 6 h of environmental hypercapnia (*C* and *D*). Hypercapnia caused a reduction in apical mitochondrion-rich cell fractional area (i.e., MRC fractional area) and a change in the appearance of the MRCs. *A* and *D* are high magnifications of the cells marked with white arrows in *B* and *C*, respectively. Bars: 2 μm in *A* and *D*; 20 μm in *B* and *C*.

However, one ultrastructural result contradicted this model: a metabolic acidosis was correlated with a 135% increase in MRCFA of rainbow trout. One explanation raised for this conflicting result was the possibility of two or more MRC subpopulations, one with apical Cl^-/HCO_3^- exchangers and one with apical proton pumps, which is consistent with studies that have localised V-ATPase in MRCs and PVCs.

Therefore, although more work is needed to confirm the specific roles of MRCs and PVCs, opposing changes in the location of acid and base transporters (via altered apical membrane morphologies) may be important mechanisms for regulating rates of net acid excretion from the gills of freshwater teleosts.

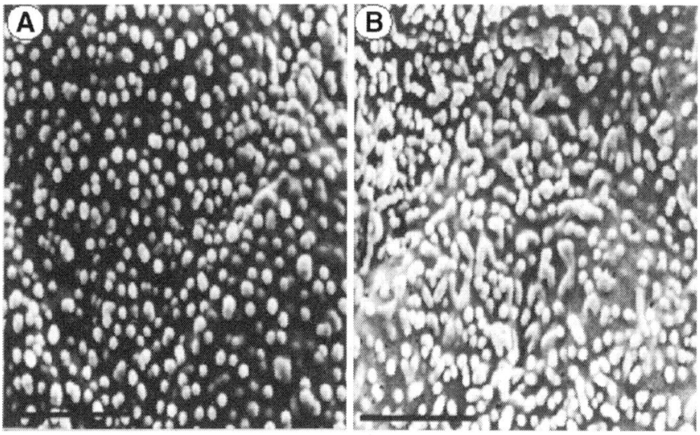

Figure: Representative high-magnification scanning electron micrographs of pavement cells (PVCs) on the filamental epithelium of brown bullhead catfish (*Ictalurus nebulosus*) during normocapnia (*A*) and after 6 h of environmental hypercapnia (*B*). Hypercapnia caused an increase in microvilli density. Bars: 2 μm.

It is important to note that with the exception of tilapia, all of the immunological evidence for apical V-ATPase is limited to salmonids, and therefore, it is premature to propose that V-ATPase is the sole route of acid secretion in all freshwater fishes. Moreover, with the exception of European flounder, all of the evidence against a direct coupling between Na^+ and H^+ are from salmonids, and therefore, it is premature to propose that apical NHEs do not function in any freshwater fishes.

In fact, recent studies on killifish, Osorezan dace, and Atlantic stingrays suggest that V-ATPase is not an important route of apical acid secretion in all fishes. Moreover, a homologous antibody for V-ATPase was used to exclusively localise the proton pump to the basolateral membranes of Na^+-K^+-ATPase-rich MRCs in the gills of freshwater acclimated killifish. Although mechanisms of acid-base transport were not proposed by the authors, it is clear that V-ATPase cannot be the direct route of acid secretion from this species. Previous pharmacological studies on killifish demonstrated a strong link between Na^+ absorption and net acid excretion when 0.1 mM amiloride was applied, which is more consistent with an apical NHE than V-ATPase and an ENaC-like channel.

In a molecular and immunological study by Hirata et al., control Osorezan dace were shown to have low levels of V-ATPase B subunit expression, which increased only slightly when exposed to low pH water (3.5) for up to 5 days. In light of much larger increases in other acid transporters, it was suggested that V-ATPase does not play a large role in acclimation to low pH (a condition that causes metabolic acidosis), and an alternative model was presented.

Finally, V-ATPase may have roles in base secretion and Cl^- absorption, as opposed to acid secretion, in nonteleost fishes. As mentioned in section V*B1*, V-ATPase has been localised to the basolateral region of MRCs that contain apical immunoreactivity for pendrin in fresh and seawater elasmobranchs. Similarly, V-ATPase is basolateral in type B intercalated cells of mammalian renal tubule collecting ducts and therefore provides a route of acid exit from the serosal side of the cells, in addition to generating base for pendrin (the apical base secreting anion exchanger). This function is also presumed for V-ATPase and pendrin-rich cells of elasmobranchs.

NHE

Intuitively, NHEs might be proposed as the dominant route of acid secretion fromfishes in seawater, where the high Na^+ concentration favours Na^+ entry across the apical membrane. HMA (0.1 mM) was shown to inhibit 75% of net acid excretion from longhorn sculpin in brackish water, suggesting that NHEs are responsible for the majority of acid secretion from this seawater teleost. This result and the finding of reduced or absent immunoreactivity for V-ATPase in seawater rainbow trout and coho salmon have led the general idea that NHEs are the proteins responsible for acid secretion at the apical side of seawater fish gills. Unfortunately, the effects of specific NHE inhibitors on acid secretion from seawater fishes have only been reported for longhorn sculpin, so it is not known if this generality can be applied to all seawater fishes.

Despite the relatively few pharmacological data that support a direct role in acid excretion, NHEs are clearly expressed in the gills of many fishes, and recent studies have demonstrated large increases in expression during treatments that cause acidosis. For example, full-length transcripts homologous to NHE2 have been cloned from longhorn sculpin, killifish, and spiny dogfish, and a full-length transcript homologous to NHE3 has been reported from freshwater Osorezan dace.

In addition to their previously mentioned roles in Na^+ absorption, these isoforms are responsible for acid secretion from mammalian intestinal and renal proximal tubule cells, and therefore are predicted to facilitate Na^+/H^+ exchange at the apical membrane of seawater fishes. Before the availability of these full-length fish sequences, heterologous antibodies were used to localise NHE2 and -3 immunoreactivity in many species of seawater teleosts and elasmobranchs.

In general, the immunoreactivity was located in MRCs, and in some cases colocalised with Na^+-K^+-ATPase. This is consistent with models that predict that basolateral Na^+-K^+-ATPase creates a low intracellular $[Na^+]$ to drive apical Na^+/H^+ exchange.

On the other hand, a homologous NHE antibody was made recently for NHE2 of the spiny dogfish, and NHE2 immunoreactivity was localised to a population of MRCs that were not rich in Na^+-K^+-ATPase. The membrane location of this shark NHE2 was not obvious in light or confocal micrographs, so its function(s) in this cell type remain(s) to be determined. Future immunohistochemcal studies with homologous antibodies to the recently sequenced NHEs will help determine if these localisation discrepancies are due to isoform, antibody, and/or species differences.

The gill of hagfish also has been shown to express an NHE, but a 1,047-bp transcript was found to be only 38–50% homologous to NHEs in GenBank. Despite this low homology, a hydropathy plot of the hagfish NHE suggested a similar topology to mammalian NHEs. In addition, relative RT-PCR was used to demonstrate an increase in expression following an injection of H_2SO_4, a treatment that causes metabolic acidosis. The increased expression was highest 2 h after the injection, a time that is consistent with the maximum rates of net-acid excretion, suggesting a direct role for this putative agnathan NHE in systemic acid-base regulation.

A recent study used homologous antibodies and molecular probes to support the hypothesis that NHE3 is the route of acid secretion from a unique freshwater teleost, the Osorezan dace. This is a freshwater teleost that lives in the extremely acidic (pH 3.4–3.8) Lake Osorezan of Japan. In this acid-tolerant fish, NHE3 immunoreactivity was localised to the apical side of MRCs that also expressed CAII, Na^+-K^+-ATPase, and NBC1.

Northern blots were also used to demonstrate that the mRNA expression of NHE and other transport related enzymes (CAII and NBC1) increased greatly when dace were transferred from water of neutral to highly acidic pH (3.5), a condition that causes metabolic acidosis. Interestingly, NHE3 is responsible for most of the acid secretion from brush-border membranes of mammalian renal proximal tubule cells where its expression also increases during metabolic acidosis. Mammalian renal proximal tubules also express CAII, Na^+-K^+-ATPase, and NBC1, making a compelling argument that the dace NHE3-rich cells function in a similar manner.

The most interesting characteristic of this study is that it suggests that an NHE-dependent mechanism may be functioning in fresh waters with a Na^+ concentration of ~1 mM and a pH below 4.0, concentrations that should make an apical NHE thermodynamically unfavourable. Although CAII, Na^+-K^+-ATPase, and NBC1 activity in the same cells would be expected to help make H^+ and Na^+ gradients more favourable by lowering intracellular [Na^+] and pH, it is hard to imagine how a >1,000-fold uphill H^+ gradient could be overcome. However, two other strategies might be involved: *1*) the expression of glutamine dehydrogenase (GDH) increased markedly in all tissues in acidic water, and *2*) the NHE3-rich cells formed a follicle in acidic water. GDH is a mitochondrial enzyme that catalyses NH_4^+ and HCO_3^- production from the substrate glutamine, and the increase in its expression suggests large increases in NH_4^+ and HCO_3^- synthesis in low pH water.

Presumably, the increased NH_4^+ synthesis would be secreted from the gills, leaving a net base HCO_3^- gain in the fish. In mammals, acidosis causes this to happen predominantly in the kidneys, where NH_4^+ is secreted by NHE3 in the proximal tubule. Therefore, as suggested by Hirata et al., the large proton gradient in dace may be overcome by secreting NH_4^+ via NHE3, instead of H^+; a demonstration of higher affinity for NH_4^+ than H^+ would help support this model. Alternatively, ammonia may leave the gills in its basic gas form (NH_3) by simple diffusion, as it does in most fishes. This would buffer the water adjacent to the gills and therefore decrease the proton gradient inhibiting acid secretion. The follicular structure might help by partially sealing the NHE3-rich cells from the low pH of the bulk water and preventing immediate loss of buffers from its lumen. Clearly, more studies like these are needed with fish-specific immunological and molecular probes to determine if NHE3 and/or NHE2 expression are regulated during acid-base disturbances in other fishes, and where they are located in the gills.

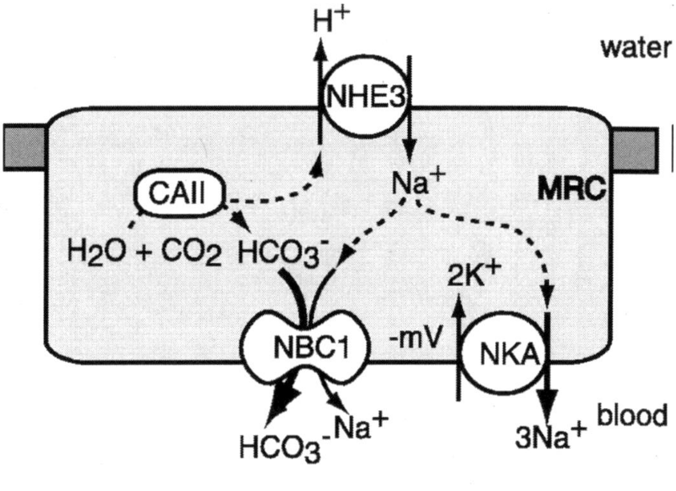

Figure: Model of acid secretion and Na^+ absorptive mechanisms in gill MRCs of FW Osorezan dace. Expression of transcripts for all four proteins were shown to increase during exposure to acidic water (pH 3.5) and were localised to a single cell type with dace-specific antibodies. In the model, acid secretion and Na^+ absorption are initiated by Na^+-K^+-ATPase, which produces a low intracellular [Na^+] and a negative inside membrane potential. These conditions then favour Na^+ absorption, in exchange for acid secretion through an apical NHE3, which increases the intracellular pH. The higher pH increases intracellular [HCO_3^-] via CO_2 hydration by carbonic anhydrase II. Finally, the increased intracellular [HCO_3^-] and negative potential drive electrogenic efflux of Na^+ and HCO_3^- across the basolateral membrane through NBC1. Electrogenic transport is indicated with unequal arrow weights. Solid arrows indicate facilitated transport, and broken arrows indicate diffusion.

H^+-K^+-ATPase

Although NHEs and V-ATPase are apical acid transporters in fish gills (and most models predict that one or the other functions in a given species), it is possible that other acid transporters also may be expressed in the gill epithelium. For instance, mammalian renal epithelial cells have H^+-K^+-ATPases, as well as NHEs, V-ATPases, along their luminal membranes. These transporters are divided further into isoforms that are thought to function in different nephron segments and may be stimulated by different pH and/or electrolyte imbalances. It has recently been demonstrated that a transcript homologous to an H^+-K^+-ATPase is expressed in the gills of elasmobranchs.

RT-PCR was used with mRNA from spiny dogfish and Atlantic stingrays to obtain sequences that are over 80% identical to HKá1 sequences at the amino acid level. In addition, two antibodies made for different parts of mammalian HKá1 stain the gills of marine Atlantic stingrays. However, the subcellular location of this staining is not clear, and RT-PCR comparisons of expression between control and hypercapnic stingrays failed to detect a difference, making further expression and localisation studies necessary to understand HKá1's role in gill acid secretion.

NBC

Base efflux across the basolateral membrane must compliment acid secretion at the apical membrane for net systemic acid excretion to occur in epithelial cells. The complete cDNA sequence of a homolog of a base transporter, NBC1, recently has been reported from the gills of rainbow trout and Osorezan dace, and a partial sequence from the gill of Atlantic stingrays is also known (Choe and Evans, GenBank accession no. AY652419).

NBC1 facilitates Na^+ and HCO_3^- efflux across the basolateral membrane of mammalian proximal tubules in a 1:3 Na^+-to-HCO_3^- ratio and has been hypothesized to have a similar function in fish gills. In support of this proposed role in systemic acid excretion, the dace NBC1 homolog was localised to the basolateral membranes of Na^+-K^+-ATPase-rich MRCs and its expression increased during exposure to low pH water.

Similarly, expression of the rainbow trout gill NBC1 homolog increased during hypercapnia. As discussed by Perry et al., it is also possible that the putative fish NBC1 homologs function in an influx mode, as occurs in mammalian pancreas cells. This would alkalise cells and would still be consistent with an increased expression during acidosis as a mechanism of intracellular pH regulation. However, at least for Osorezan dace, NBC1 was specifically localised in MRCs that were also shown to have apical NHE3 and basolateral Na^+-K^+-ATPase. Because the activities of these two highly expressed transporters would alkalise and hyperpolarise the cell, it is more likely that dace NBC1 functions in an efflux mode. Further studies are clearly needed to determine the role of NBC1 in fish gills.

Base Secretion

As mentioned above, AE1 has been localised to the apical side of a subpopulation of Na^+-K^+-ATPase-rich MRCs in the gills of two freshwater teleosts (coho salmon and tilapia) with an antibody raised against the rainbow trout erythrocyte band 3 protein. Interestingly, this finding of two or more subpopulations of freshwater MRCs agrees with some of the previously mentioned morphological and immunolocalisation results for freshwater salmonids and suggests that there may be base- and acid-secreting freshwater MRCs. Simultaneous localisation of AE1, V-ATPase, and Na^+-K^+-ATPase in the same sections would be needed to verify this possibility. Only one study has examined the effects of an acid-base stress on the expression of AE1 in the gills. This study demonstrated an increased number of cells expressing AE1 during chronic $NaHCO_3$ infusions (rainbow trout), which suggests a role for this exchanger in base secretion. Although AE1 is predicted to also secrete base in seawater teleosts, evidence is still lacking. Wilson et al. demonstrated a loss of immunoreactivity for AE1 when coho salmon were acclimated from fresh water to seawater. This could mean that a different base transporter was activated or that AE1 expression was diminished below detectable levels (but still functional with the higher external [Cl^-]). Clearly, more immunological and molecular data are needed before the model of apical AE1 in Na^+-K^+-ATPase-rich MRCs is applied to all freshwater and seawater teleosts.

The use of AE1 in the apical membrane of Na^+-K^+-ATPase-rich cells to secrete base and absorb Cl^- in the fish gill is different from intercalated cells of mammalian renal collecting ducts in a two important ways. For example, in B type cells, the recently discovered anion exchanger pendrin is thought to be the apical acid exchanger instead of AE1, and basolateral V-ATPase is the route of acid exit across the basolateral membrane. Alternatively, with the exception of killifish, V-ATPase is only associated with the apical membrane in freshwater teleosts and therefore is probably not the basolateral route of acid exit. It is still unclear how acid exits the basolateral membrane of the presumptive teleost base-secreting cells, and what role, if any, basolateral Na^+-K^+-ATPase plays in apical Cl^-/HCO_3^- exchange.

The recently developed model of base secretion and Cl^- absorption for the elasmobranch gill epithelium is much more similar to mammalian renal type B intercalated cells than the teleost model. For example, apical pendrin and basolateral V-ATPase immunoreactivity occur in a subpopulation of MRCs.

Although immunoreactivity for pendrin and V-ATPase was inversely correlated with acclimation salinity, suggesting a direct role in Cl^- absorption for ion regulation, similar experiments are needed in relation to acid-base disturbances to verify a role in base secretion. In addition, the antibody used was raised against human pendrin, and no molecular data have been obtained for this putative anion exchanger in elasmobranchs. Therefore, molecular sequence for this putative exchanger is needed to determine its evolutionary affiliation within the anion exchanger families.

Nitrogen Balance

Unlike biomolecules such as fatty acids and sugars, excess amino acids are not stored or excreted. Their carbon skeletons are instead converted into intermediate metabolic fuel sources after removing the á-amino group. While the carbon backbone is converted into fatty acids, ketone bodies, and carbohydrates, the excess nitrogen is usually excreted. This net gain in excess nitrogen is magnified in the high-protein diets of many fishes where most of their dietary carbon is extracted from amino acids.

In addition, fasting fishes can use their own muscle proteins as a source of amino acids, which are deaminated to produce ATP for maintenance, or converted to glucose by the liver. Excess nitrogen is not usually stored in tissues because it is either too toxic (i.e., ammonia) or too energetically costly to convert to less toxic forms (i.e., uric acid and urea). For example, ammonia, the most useful form of nitrogen for synthesizing amino acids, is highly toxic and therefore cannot be sequestered at any significant level. Although less toxic than ammonia, urea and uric acid require ATP consumption for their synthesis, counteracting the benefits of catabolising proteins for energy. In addition, uric acid precipitates out of solution and is produced by terrestrial animals that need to conserve water (e.g., birds and reptiles), obviously not a problem for at least freshwater fishes. Somefishes (elasmobranchs and coelacanths) do synthesize and sequester nitrogen as urea but for osmoregulation, not nitrogen storage. In fact, eukaryotes lack ureases, and therefore chondrichthian fishes cannot reincorporate the urea they synthesize unless they harbour urease-expressing bacteria. Although an apparently successful strategy (>1,000 extant species), the high urea concentrations of ureosmotic fishes (>300 mM) are toxic and must be balanced with stabilising molecules such as trimethylamine oxide. Therefore, net production of nitrogenous waste occurs in most fishes, under most conditions, and because of toxicity and energetic costs of storage, most fishes deal with waste nitrogen by excreting it immediately. Nitrogenous molecules other than ammonia and urea are excreted by fishes (e.g., taurine, creatine, creatinine, purines, and methylamines), but their quantitative contributions are small and mechanisms of excretion poorly understood. Mechanisms of ammonia and urea metabolism and excretion have been studied extensively in fishes, and several excellent reviews are available. Here, we focus on mechanisms of excretion by the gills.

The gills are the primary site of ammonia and urea excretion from fishes for many of the same reasons that they are the primary site of gas exchange and systemic ion and acid-base transport (e.g., large surface area, perfusion by 100% of cardiac output, large ventilation rates, small diffusion distances, and contact with a voluminous mucosal medium). Branchial and renal nitrogen excretion rates have been compared in several species, and >80% of the total nitrogen (ammonia plus urea) was excreted from the gills of most species. The vast majority of fishes, including almost all actinopterygians and agnathans, excrete the majority of their nitrogenous waste as ammonia and are referred to as ammonotelic. This appears to be the default condition for fishes, reflecting the ease with which NH_3 can diffuse through tissues and into a large external aqueous environment. Alternatively, coelacanths, most elasmobranchs, and a few teleosts excrete most of their nitrogenous waste as urea and are referred to as ureotelic. Ureotelism appears to have evolved either as a mechanism of detoxifying ammonia in environments that inhibit ammonia excretion (e.g., air exposure and high pH) or as a mechanism of osmoregulation in seawater (coelacanths and elasmobranchs).

Ammonia

Ammonia is a weak base and occurs as both a gas, ammonia (NH_3), and an ion, ammonium (NH_4^+), in aqueous solutions; the sum of both forms is referred to as total ammonia (T_{amm}). Because the pH of fish blood and intracellular fluid is over one unit below the pK of ammonia (~9–10), ~95% of T_{amm} exists as NH_4^+ infish tissues. Although the specific mechanisms of T_{amm} toxicity in animals are not completely understood, high micromolar concentrations are clearly incompatible with many tissue functions. The most acute effects of ammonia are probably related to the ability of its ionic form, NH_4^+, to substitute for K^+ in ion transporters and disrupt electrochemical gradients in central nervous systems. This may explain why fishes, which have less advanced central nervous systems, are more tolerant of T_{amm} than mammals. For instance, T_{amm} concentrations in human blood are kept below 40 μM by hepatic urea synthesis, and even small increases (50%) cause neural pathologies. Control T_{amm} concentrations reported for fish blood are up to an order of magnitude higher and more variable (between species and study). Wood attributed much of the variability to different blood collection and analysis techniques and estimated that the "true" control arterial T_{amm} concentrations offishes are <500 μM and typically between 100 and 200 μM. Although fishes are more tolerant of T_{amm} than mammals, they are susceptible to the same toxic effects at higher concentrations and therefore must excrete T_{amm} at the same rate it is synthesized.

Unlike in mammals, where ammonia is synthesized at the site of excretion, i.e., kidneys, <25% of excreted ammonia is synthesized at the gills of fishes. Most of the ammonia excreted by the gills is produced in other tissues and cleared from the blood. During control conditions, the liver appears to synthesize the majority of ammonia in fishes, with skeletal muscle, kidneys, and gills contributing in descending quantitative order. However, skeletal muscle can dominate ammonia synthesis during glycolytic exercise or hypoxia via adenylate deamination. Whole body T_{amm} excretion rates have been measured for many species, including teleosts, elasmobranchs, and agnathans. Under control conditions, most ammonotelic fishes have excretion rates between 100 and 350 μmol $\cdot$ kg^{-1} $\cdot$ h^{-1}, and ureotelic elasmobranchs have excretion rates below 50 μmol $\cdot$kg^{-1} $\cdot$h^{-1}. However, T_{amm} excretion rates over 1,000 μmol $\cdot$ kg^{-1} $\cdot$ h^{-1} have been measured after feeding and NH_4Cl infusion, demonstrating a high capacity for T_{amm} transport across the gill epithelium.

Despite being studied exhaustively over the past 30 years, the exact mechanisms of ammonia transport across the branchial epithelium of fishes remain somewhat controversial, because it is difficult to measure and/or control transepithelial gradients of pH, P_{NH3}, NH_4^+, and electrical potential simultaneously in wholefish or whole tissue experiments. Although at least five potential mechanisms have been hypothesized, with some data supporting each, the majority of current experimental data support three pathways of ammonia excretion from the gills of fishes. These include *1*) NH_3 diffusion, *2*) NH_3 diffusion trapping, *3*) Na^+/NH_4^+ exchange, and *4*) NH_4^+d ffusion. Each mechanism is not exclusive, and their quantitative contributions are determined by conditions such as external water, salinity, pH, and buffer capacity.

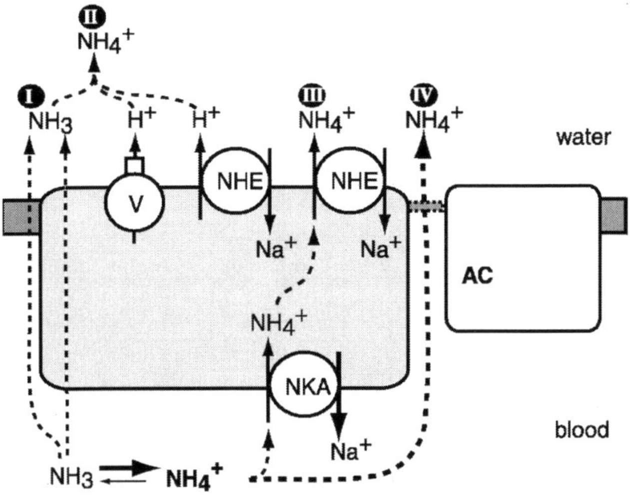

General Ammonia Secretion

FIG. 35. Composite model of ammonia secretion pathways in the gills of fishes (e.g., agnathans, elasmobranchs, and teleosts), numbered in decreasing order of occurrence (Roman numerals). *I*, diffusion of NH_3 either through or between epithelial cells down favourable blood-to-water diffusion gradients. *II*, diffusion of NH_3 that relies on proton secretion to "trap" ammonia as NH_4^+. *III*, transport of ammonia through NHE and/or Na^+-K^+-ATPase. *IV*, diffusion of NH_4^+through the leaky junctions that occur between MRC and AC in seawater teleosts. *Pathway I* can occur for any cell type, *pathways II* and *III* can only occur in acid secretion cells, and*pathway IV* can only occur with salt secretory cells of teleosts.

NH_3 Diffusion

Experiments on an agnathan, an elasmobranch, and a number of freshwater teleost species suggest that most T_{amm} crosses the branchial epithelium as NH_3, down favourable blood-to-water diffusion gradients (teleost data reviewed extensively by). Three lines of evidence support this conclusion. *1*) Moderate decreases in external water pH, which decreases $[NH_3]$ in the external water via protonation, increased the rate of T_{amm} excretion from freshwater rainbow trout, carp, and goldfish. *2*) Increases in external water pH, which increases $[NH_3]$ in the external water via deprotonation, decreased the rate of T_{amm} from freshwater rainbow trout. *3*) Ammonium injections $[NH_4Cl$ or $(NH_4)_2SO_4]$ caused an initial increase in T_{amm} excretion, in excess of net acid excretion, from Atlantic hagfish, spiny dogfish sharks, rainbow trout, and channel catfish, suggesting that at least some of the infused ammonium dissociated and crossed the gills as NH_3. These ammonium injections also caused a metabolic acidosis, which presumably reflects residual H^+, left in the blood after dissociation of NH_4^+ and excretion of NH_3.

NH_3 Diffusion Trapping versus Na^+/NH_4^+ Exchange

V*B1* and VI*D*, early models of gill Na^+ absorption suggested the presence of a Na^+/NH_4^+ exchange system in freshwater teleosts. This hypothesis was formed from observations that amiloride addition to and Na^+removal from external water decreased net ammonia excretion rates and that ammonium salt injections increased Na^+ absorption in freshwater teleosts. However, as pointed out by Avella and Bornancin, all these results can also be explained by diffusion trapping of NH_3 via H^+ or CO_2secretion into an unstirred layer of water on the apical surface of the gills. For instance, amiloride addition and Na^+ removal could inhibit T_{amm}excretion by inhibiting acid secretion from either NHEs or the V-ATPase/ENaC-like system. A test of this hypothesis was conducted on rainbow trout by adding HEPES to the water, which should make diffusion trapping of NH_3 negligible, because the external buffer would bind with any secreted acid at the apical surface of the gills, minimising acidification of the unstirred layer. Despite a 90% inhibition of Na^+ uptake by amiloride in the HEPES-buffered fresh water, T_{amm} excretion was unaffected. Similar results were observed for Lahontan cutthroat trout (*Oncorhychus clarki henshawi*), which live in highly buffered, alkaline water of Pyramid lake. This lack of Na^+/NH_4^+exchange in freshwater trout probably reflects the use of apical V-ATPase and an ENaC-like channel for acid secretion and Na^+ absorption (instead of apical NHEs), as is hypothesized for freshwater salmonids, tilapia, and flounder.

Most seawater and some freshwater fishes, however, are thought to use apical NHEs for acid secretion and Na^+ absorption, and therefore at least have the potential for apical Na^+/NH_4^+ exchange. Nevertheless, most of the data for seawater fishes under normal, low external T_{amm} conditions suggest that little to no ammonia excretion is via Na^+/NH_4^+ exchange.

For example, Na^+ removal from and amiloride addition to external water inhibited net T_{amm} excretion from intact spiny dogfish sharks and Gulf toadfish (*Opsanus beta*) by <50%. In addition, similar treatments had no effect on net T_{amm} excretion from intact Atlantic hagfish and little skates, and amiloride, in the presence of oubain, had no effect on net T_{amm} excretion from isolated perfused heads of spiny dogfish sharks and toadfish. Therefore, under normal conditions when the T_{amm} in external water is low, diffusion (NH_3 and/or NH_4^+) probably dominates. However, even if we assume that seawater fishes have apical NHEs that can substitute NH_4^+ for H^+, Na^+/NH_4^+ exchange might not be needed in most cases because of favourable diffusion gradients for NH_3 and NH_4^+. Alternatively, in crowded or confined waters, the T_{amm} can rise rapidly, eliminating the favourable diffusion gradients for NH_3 and NH_4^+, requiring active secretion of ammonia against a gradient.

For fishes, the only unequivocal demonstration of T_{amm} excretion against a gradient is from the giant mudskipper, an amphibious, air-breathing teleost that lives in brackish-water swamps of Southeast Asia. Mudskippers regularly leave water and have apparently evolved a mechanism of active ammonia excretion in the absence of favourable diffusion gradients. Randall et al. demonstrated that T_{amm} excretion from this species was maintained in water of 30 mM T_{amm}, without any increase in blood concentration from its normal value of ~150 μM. T_{amm} excretion was not affected by HEPES, suggesting that diffusion trapping via boundary layer acidification did not play an important role in ammonia secretion. Alternatively, it was inhibited by amiloride and ouabain, suggesting the use of NHEs and Na^+-K^+-ATPase for active NH_4^+ secretion. Wilson et al. used heterologous antibodies to demonstrate immunolocalisation for NHE2 and NHE3 in the apical area of MRCs that also expressed high levels of basolateral Na^+-K^+-ATPase. Therefore, it was hypothesized that active branchial ammonia excretion from these mudskippers takes place by a mechanism very similar to mammalian renal proximal tubules, in which NH_4^+ is secreted across the apical membrane by substituting for H^+ in NHE. Na^+-K^+-ATPase in the basolateral membrane assists by creating a low intracellular $[Na^+]$ that drives NHE, and by directly facilitating NH_4^+ entry into the cell (via substitution for K^+ on Na^+-K^+-ATPase). More recent experiments suggest that the gills of this species also secrete acid to trap any excreted NH_3 as NH_4^+ and maintain low skin NH_3 permeability by a combination of high cholesterol and fatty acid content.

As discussed above, freshwater Osorezan dace may also be capable of apical Na^+/NH_4^+ exchange to allow acid secretion and Na^+ absorption in a highly acidic environment. In addition, hagfish, elasmobranchs, and marine teleosts are all thought to have a similar arrangement of transporters for acid secretion in MRCs, and therefore also may be able to secrete ammonia actively when external T_{amm} increases so that diffusion gradients are reversed from normal. Future studies should measure expression levels of these potential NH_4^+ transporters during high external ammonia to test this possibility. In addition, heterologous expression systems should be used to determine if the recently cloned fish NHE2- and NHE3-like transcripts can transport NH_4^+ when it is present at physiological concentrations.

NH_4^+ Diffusion

NH_4^+ diffusion is probably minimal in all fishes except seawater teleosts and lampreys, because of low permeability to cations by the branchial epithelium. Seawater teleosts and lampreys have shallow tight junctions between MRCs, which increase cation permeability for Na^+ secretion and therefore are probably more permeable to NH_4^+ than other fishes. This was suggested by experiments on seawater longhorn sculpin, which did not develop a blood alkalosis when exposed to high environmental T_{amm} concentrations, suggesting that ammonia loading occurred as the acid-base neutral form, NH_4^+, and not the basic form, NH_3. In addition, Goldstein et al. demonstrated that net ammonia excretion rates from seawater longhorn sculpin and toadfish were sensitive to changes in external NH_4^+ concentrations but not external NH_3 concentrations.

Urea

Blood urea concentrations have been measured in many fish species and appear to vary by taxonomic group and by conditions that inhibit passive ammonia excretion via diffusion. Ammonotelic teleosts and agnathans, which include the majority of their respective groups, typically have blood plasma urea concentrations below 2–3 mM, reflecting their low urea synthesis rates and dependence on ammonia excretion to remove nitrogenous waste.

Higher plasma urea concentrations have been measured for teleosts that live in environments that are unfavourable for ammonia excretion. For example, the Lahontan cutthroat trout of Pyramid Lake, a cyprinid of Lake Van, Turkey (*Chalcalburnus tarichi*), and a tilapia of Lake Magadi, Kenya (*Alcolapia grahami*) have plasma urea concentrations of 8.15, 36.18, and 10.52 mM, respectively.

These species all live in waters with a pH between 9.4 and 10.0 that inhibit passive NH_3 excretion via diffusion trapping and appear to elevate urea synthesis to partially (Lahontan cutthroat and cyprinid), or completely (Lake Magadi tilapia), compensate for reduced ammonia excretion. A few other fishes have elevated plasma urea concentrations, either during stress or confinement (gulf toadfish) or air exposure (e.g., lungfishes). By far, the highest blood urea concentrations are found in fishes that are ureosmotic, which include coelacanths and almost all elasmobranchs. The only species of the latter group known to be ammonotelic belong to a single family of stenohaline freshwater stingrays in South America, the Potamotrygonidae.

As with ammonia, the majority of urea appears to be synthesized in the liver offishes, regardless of the enzymatic pathways used to synthesize it. Arginase and uricolytic enzymes synthesize the majority of urea in ammonotelic fishes and appear to be present in most species. A fully functional ornithine-urea cycle (OUC) appears only in chondrichthyes, coelacanths, and a limited number of teleosts. Because the majority of teleosts are ammonotelic, few studies have investigated mechanisms of urea transport in the gills of teleosts until the past decade. These have largely focused on two species that can be ureotelic and express a fully functional OUC, the gulf toadfish and the Lake Magadi tilapia.

Pulsatile Urea Excretion

Gulf toadfish are normally ammonotelic when they are first acquired from the wild, but soon become facultatively ureotelic (>70% of nitrogen excreted as urea) when crowded or confined in aquaria. Under these conditions, gulf toadfish excrete urea from the gills in discrete pulses that last under 3 h each and occur about one time per day. Remarkably, urea excretion rates rise from near zero (~10 μmol · kg^{-1} · h^{-1}) during steady-state, nonpulse times to over 300 μmol · kg^{-1} · h^{-1} during pulses, and blood concentrations decrease by ~1 mM. This excretion pulse is not due to rapidly increased blood urea concentrations or urea synthesis rates, because neither increases sharply before or during a pulse event. A series of in vivo experiments ruled out two of three hypothetical excretion mechanisms that could explain the observed pulses: *1*) involvement of a normally active back-transport system similar to elasmobranchs and *2*) changes in the general permeability of the gills.

For example, neither Na^+ removal from nor urea analogue addition to the external water increased urea excretion from toadfish during nonpulse steady-state periods, suggesting the absence of a back-transport system as occurs in elasmobranchs. Such treatments would be expected to inhibit a back-transport system and increase urea excretion. A general increase in gill perfusion or permeability was also ruled out by experiments that demonstrated a lack of increased urea excretion after injection of a strong branchial dilator (L-isoprenaline) and by experiments that showed no change in the permeabilities of polyethylene glycol-4000 or water during urea pulses.

Further experiments suggested that the urea permeability could be regulated by periodic translation, insertion, and/or activation of specific urea transporters. For instance, elevating urea concentrations in the external water to three times the blood concentration (30 mM) caused absorption of urea into the blood at each pulse event that was proportional to the relative concentrations in the water and blood, demonstrating that the transport system is reversible. This is a characteristic of rat inner medullary collecting ducts, which express the facilitated urea carrier UT-2.

Walsh et al. now have cloned, sequenced, and functionally expressed a urea transporter from toadfish gills (tUT), which was the first urea transporter sequenced from fish gills. The 1,814-bp cDNA (GenBank accession no.165893) codes for a protein that is over 55% homologous to mammalian, amphibian, and elasmobranch kidney urea transporters and exhibits phloretin-sensitive urea transport when expressed in oocytes. Interestingly, of six tissues assayed with Northern blots including the kidneys, tUT expression was only detected in the gills of toadfish, confirming earlier divided chamber, and pharmacological studies that suggested that gills are the primary site of urea excretion.

Therefore, it was concluded that tUT is the protein responsible for transient urea permeability in toadfish gills. However, no statistical change in tUT expression levels were detected with Northern blots during urea pulses, suggesting that regulation of urea excretion is posttranscriptional. In support of this hypothesis, Laurent et al. used transmission electron micrographs to demonstrate an increase in vesicle numbers and proximity of vesicles to the apical membranes of PVCs in urea-pulsing toadfish. The authors suggested that tUT might be expressed in these vesicles, which are delivered to the apical membrane during a urea pulse.

Urea Excretion at High External pH

The Lake Magadi tilapia is another unique teleost with respect to nitrogen metabolism. As mentioned earlier, it lives in the highly basic (pH 10.0) and buffered (CO_2 content = 180 mM) waters of Lake Magadi in Kenya that originate from volcanic hot springs. The Lake Magadi tilapia thrives in these conditions by having a fully functional OUC and excreting essentially all of its waste nitrogen as urea through the gill, which has an exceptionally high urea permeability (4.74×10^{-5} $cm^{-1} \cdot s^{-1}$) (636). Furthermore, the nitrogen excretion rates (~3,302 $\mu mol \cdot kg^{-1} \cdot h^{-1}$) are high for teleosts, presumably because of high temperatures in the lake and the high nitrogen content of their cyanobacteria diet.

The results of in vivo, molecular, and microscopy experiments suggest that the gills of Lake Magadi tilapia excrete urea by a mechanism similar to the gulf toadfish gills during a pulsing event. For instance, the ratio of thiourea to urea permeability was 0.18, and urea transport was shown to be reversible and proportional to the relative concentrations in the external water and blood plasma, demonstrating a bidirectional transport system. These are both characteristics of toadfish gills during urea pulses, and of the rat inner medullary collecting duct. In addition, a 1,699-bp cDNA (GenBank accession no.AF278537) coding for a 475-amino acid protein (mtUT) was cloned, sequenced, and functionally expressed from Lake Magadi tilapia. It is 75% identical to the tUT protein and is thought to be responsible for the exceptionally high urea permeability of the gills in this species. Finally, vesicles structurally similar to those proposed to contain tUT in toadfish were observed near the apical membranes of PVCs in Lake Magadi tilapia gill and were likewise proposed to contain mtUT. Clearly, immunolocalisation and in situ hybridisations are needed to further test this hypothesis, especially in light of the more recent immunolocalisation of a homologous UT in the MRC of eels. One important difference between urea excretion from Lake Magadi tilapia and toadfish is the lack of pulses in the former. Interestingly, although the amino acid sequences of the mtUT and tUT proteins are similar, they do differ with respect to potential PKC phosphorylation sites at their carboxy termini. Walsh et al. have speculated that this could potentially explain the differences in how urea excretion is regulated between the two species.

Urea Transporters in Ammonotelic Fishes

It was assumed that the relatively small amounts of urea excreted by ammonotelic teleosts crossed the gills through nonspecific pathways, such as via diffusion through lipid membranes. However, when a radiolabelled probe for tUT was hybridised with RNA from 14 species of marine teleosts under low-stringency conditions, homologous transcripts were detected in almost all of them. This suggests that specific facilitated urea transporters are expressed in the gills of most, if not all, teleosts regardless if they are ureotelic or not. In support of this hypothesis, Mistry et al. cloned and sequenced a urea transporter from the ammonotelic Japanese eel (eUT) that is 486 amino acids and used a homologous antibody to localise it to MRCs. The function of this branchial urea transporter in an ammonotelic teleost is uncertain but could be a way of ensuring secretion of urea that is synthesized via non-OUC pathways under control conditions (e.g., cleavage of dietary arginine, and from uricolytic pathway of purine catabolism). Interestingly, the expression of this branchial urea transporter was higher in seawater than in fresh water. The reason for higher expression in seawater is unclear but may be related to reduced urinary flow rates (and therefore reduced urea excretion by the kidneys) in high salinities requiring increased secretion by the gills.

Therefore, the expression of branchial facilitated urea transporters may be common to all teleosts but is probably most important in species that elevate urea synthesis to detoxify ammonia. The immunolocalisation of eUT in MRCs of eels is not consistent with the hypothesis that vesicles near the apical membranes of PVCs contain urea transports (tUT and mtUT) in gulf toadfish and Lake Magadi tilapia. Clearly, immunohistochemical localisation of tUT and mtUT are needed to evaluate this discrepancy.

Retention Mechanisms in Elasmobranchs

Despite being ureotelic and excreting over 80% of their nitrogenous waste as urea, elasmobranchs require branchial mechanisms to minimise loss of urea that is required to maintain osmotic balance in seawater. For instance, Pärt et al. pointed out that the urea excretion rates of elasmobranchs would be ~10,000 μmol · kg^{-1} · h^{-1} if their gills had urea permeabilities similar to teleosts, which is 40-fold greater than their actual excretion rates (250 μmol · kg^{-1} · h^{-1}). A low urea permeability by elasmobranch gills was first reported for the spiny dogfishby Boylan and later verified by Fines et al. to be 3.2×10^{-8} cm/s, which is over 60-fold less than the permeability of rainbow trout and eel gills. Unfortunately, detailed investigations of the mechanisms responsible for this low permeability have only begun in the past 10 years, and large gaps still exist in the current models. Although the low urea permeability of elasmobranch gills was originally assumed to be a purely structural adaptation, recent data suggest a combination of structural and active transport mechanisms.

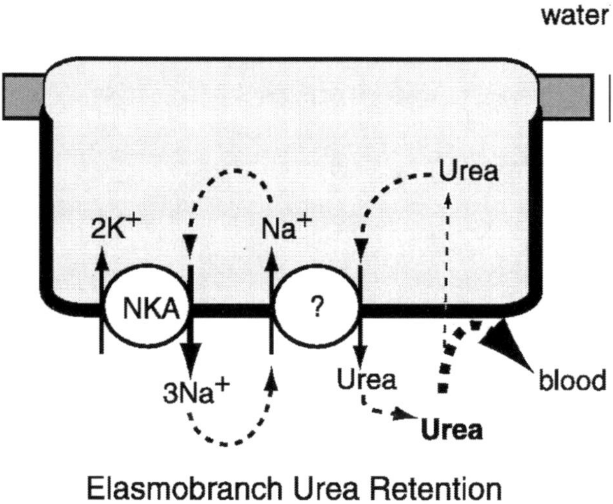

Figure: Model of urea retention mechanisms proposed for marine elasmobranchs. The basolateral membranes of spiny dogfish shark gills were shown to have a high cholesterol content, which reduces the membrane's permeability to urea (indicated by a thick line for the basolateral membrane). Although this is thought to minimise urea entry into the cells, some still leaks in and is thought to be transported back across the basolateral membrane by an unidentified, Na^+-dependent urea transporter. The Na^+ gradient required for this urea transporter is thought to be maintained by Na^+-K^+-ATPase.

The structural adaptations may be due to an unusual composition of the plasma membranes in the gill epithelium. For instance, basolateral membrane vesicles (BLMV) from spiny dogfish shark gill tissue were shown to have the highest cholesterol-to-phospholipid ratios ever recorded for a natural biological membrane. This was proposed as a mechanism to minimise urea entry into the epithelial cells across the basolateral membrane, because cholesterol is known to reduce urea permeability by allowing a tighter fit between adjacent phospholipid molecules.

Data from in vivo and isolated perfused head experiments suggested the presence of an active, back-transport system in the gill epithelium spiny dogfish sharks. For instance, urea excretion rates from this species were unaltered by up to 50% changes in urea concentrations of plasma or isolated-head perfusion fluids, suggesting a saturated transport mechanism. In addition, urea excretion rates from spiny dogfish were increased by the competitive urea analogues acetamide and thiourea and by the noncompetitive inhibitor phloretin, suggesting the presence of a carrier-mediated transport system.

Finally, a washout experiment with isolated perfused heads of spiny dogfishsharks demonstrated a 14-fold greater urea transport rate across the basolateral membrane than across the apical membrane. These data support a model of the gill epithelium as an "intermediary compartment" where the urea gradient across the apical membrane is lowered by active transport of intracellular urea across the basolateral membrane.

More recently, a detailed study of BLMV from spiny dogfish shark gills further characterised the putative carrier-mediated transporter. Urea uptake by BLMV was inhibited by the urea analogues *N*-methylurea and nitrophenolthiourea and was inhibited in a dose-dependent manner by phloretin, confirming the presence of carrier-mediated transport. Furthermore, urea uptake was stimulated by ATP, and this stimulation was sensitive to ouabain, suggesting that the urea transporter was secondarily active, dependent on Na^+ and/or K^+ gradients created by Na^+-K^+-ATPase. Other experiments showed that Na^+ gradients, and not K^+ gradients, stimulated uptake, suggesting that the urea transporter is Na^+ dependent and probably uses the Na^+gradients created by Na^+-K^+-ATPase. Finally, K_m for urea was 10.1 mM, which is much lower than the blood plasma urea concentration. The authors suggested that this may be indicative of an intracellular urea scavenging role for the transporter that actively returns urea to the blood.

Therefore, the current model for urea retention by elasmobranch gills suggests that the role of basolateral membrane is to minimise entry of urea with a low permeability via a high cholesterol-to-phospholipid ratio. Most of the urea that does enter the cells is then transported back out through a yet to be identified Na^+-stimulated urea transporter in the basolateral membrane, which maintains a low intracellular urea concentration to minimise urea diffusion across the apical membrane.

Obviously, further experiments are needed to identify and localise the putative basolateral, Na^+-stimulated urea transporters. Transcripts that are homologous to those cloned from the kidneys of elasmobranchs are expressed at low levels in the gills of spiny dogfish and Atlantic stingrays, but they are passive and not thought to be a part of the basolateral membrane back transport mechanism. Further experiments on other species are also needed to determine if the proposed basolateral membrane adaptations of spiny dogfish sharks are shared by other elasmobranchs.

Chapter 7

Neural, Hormonal and Paracrine Control

Given the complexity of the perfusion pathways in the fish gills and the location of specialised transporting cells and paracellular pathways, one might hypothesize that complex control systems have evolved to provide for homeostasis and the potential for rapid and long-term responses to environmental or internal changes. Interestingly, many of these signalling agents have both hemodynamic and ionic transport effects.

Intrinsic Control

Cholinergic Neurons

Stimulation of the "vagosympathetic" nerve trunk to the isolated third gill arch of the Atlantic cod produced an increase in branchial vascular resistance, which was reversed by the addition of atropine, corroborating studies that demonstrated that acetylcholine produced an increase in branchial resistance in the rainbow trout. The early proposition that constriction of postlamellar, efferent vessels underlies this response has been confirmed by in vivo videomicroscopy that visualised constriction of the efferent filamental vessels in the rainbow trout, which was blocked by atropine. The site of constriction is generally assumed to be the prominent sphincter in the proximal portion of the efferent filamental artery that is innervated by vagal cholinergic fibres in a variety of teleost species. Constriction in the postlamellar, efferent filamental artery is accompanied by increased blood flow into the ILV and increased flow through the lamellae, both of which are blocked by atropine. Similar, atropine-sensitive, changes in the perfusion pattern and gill resistance can be elicited by hypoxia, which would increase the functional surface area of the gills, appropriate to compensate for hypoxic conditions. Moreover, diversion of blood into the ILV may provide greater perfusion of MRCs (although no evidence exists suggesting perfusion limitation to transport) and/ or plasma skimming to enrich the hematocrit of the efferent filamental blood. Atropine sensitivity suggests a muscarinic receptor in the efferent gill vasculature, but no biochemical or molecular characterisation of the putative receptor has been published. However, an M_3 type muscarinic receptor has been characterised in the spiny dogfishventral aorta, so it is possible that this receptor mediates the cholinergic effects in the gills also.

To our knowledge only a single study has demonstrated that cholinergic innervation may have a direct effect on ionic transport across the gill epithelium. May and Degnan found that acetylcholine inhibited the I_{sc} across the isolated killifish opercular epithelium.

Adrenergic Neurons and Chromaffin Tissue

Epinephrine and norepinephrine reach gill tissues via both neurotransmission from sympathetic innervation (apparently only in teleosts), and from stimulation of chromaffin cells in the teleost and elasmobranch interrenal gland, which is associated with the head kidney. Intravenous infusion of catecholamines into a variety of teleost species is associated with a biphasic response in gill resistance: an initial, á-adrenergic-mediated increase, followed by a longer-lasting, β-adrenergic-mediated fall in resistance. Stimulation of the sympathetic chain anterior to the celiac ganglion in the cod produced the same á-mediated vs. β-mediated responses.

The -adrenergic-mediated fall in gill resistance was associated with an increased oxygen uptake and was presumed to be the result of lamellar recruitment. Accordingly, hypoxia elicits an increase in plasma catecholamine levels, with the attending changes in gill perfusion, as part of a complex response to maximise oxygen delivery to the tissues. Measurement of branchial venous outflow in the cod in vivo has demonstrated that the á-mediated increase in resistance is secondary to constriction of the arteriovenous anastomoses between the efferent filamental artery and the ILV, where nerve terminals have been observed. The β-mediated fall in branchial resistance is generally considered to be the result of dilation of afferent lamellar arterioles, which produces lamellar recruitment. It is noteworthy that catecholamines also produce preferential arterio-arterial flow in the hagfish gill pouch, suggesting an ancient origin for at least the sympathetic, branchial control mechanisms.

In addition to potential effects of catecholamines on gill permeability and transport produced by changes in perfusion patterns, it is clear that these vasoactive mediators can have direct effects on branchial ionic transport. The initial characterisation of the killifish operculum as a model for teleost gill salt extrusion demonstrated that epinephrine inhibited the I_{sc} produced by the tissue. Subsequent studies demonstrated that Cl^- fluxes across the tissue could be inhibited by á-adrenergic activation and stimulated by β-adrenergic activation and that inhibition versus stimulation of the I_{sc} was via $á_2$-adrenergic and and $β_1$-adrenergic receptors, respectively. Moreover, cAMP is the second messenger for the β-stimulation. A more recent study has demonstrated that stimulation of the trigeminal nerve associated with the opercular epithelium produced a á-mediated inhibition of the I_{sc} with an associated increase in intracellular inositol trisphosphate (IP_3) levels. Adrenergic neurons may also play a role in modulating Ca^{2+} balance, because one study demonstrated that epinephrine injection or stimulation of adrenergic branchial nerves inhibited $^{45}Ca^{2+}$ uptake into the gill tissue of the rainbow trout.

Serotonergic Neurons and Neuroepithelial Cells

The fact that the constrictory response to nerve stimulation in the cod could not be totally abolished in some preparations by pretreatment with either 10 μM atropine or phentolamine (á-adrenergic antagonist) suggested that some nonadrenergic, noncholinergic (NANC) fibres may exist in the gills. Serotonergic fibres now have been described in the gills of a variety of teleost species, and serotonin increased branchial resistance in a number of species in vivo and in vitro. Video observation of blood flow through the rainbow trout gills has shown that serotonin redistributes blood flow to the more proximal parts of the filament, by constriction of distal portions of the efferent filamental artery and possibly the afferent filamental artery. The resulting reduction of lamellar perfusion presumably accounts for the reduction in gas exchange when serotonin is infused into two species of teleosts. In the cod, serotonin also dilates the arteriovenous anastomoses, shunting more blood into the ILV. Interestingly, serotonin apparently stimulates the release of catecholamines in at least the rainbow trout. Sensitivity to methysergide suggests that the serotonin effect is mediated by 5-HT_2 receptors in at least three species of teleosts. Serotonin is also produced in gill neuroepithelial cells, which are on efferent side of gill filaments and assumed to be O_2 sensors, but the relative roles of NEC release versus neuronal release are unclear.

To our knowledge the effects of serotonin on fish gill transport are unstudied, but the nonmetabolised, serotonin agonist 8-hydroxy-2-(di-*n*-propylamino)tetraline (8-OH-DPAT) stimulates Na^+-K^+-ATPase expression and enzymatic activity in the opossum proximal tubule (61). Therefore, similar studies in fish gills would be interesting.

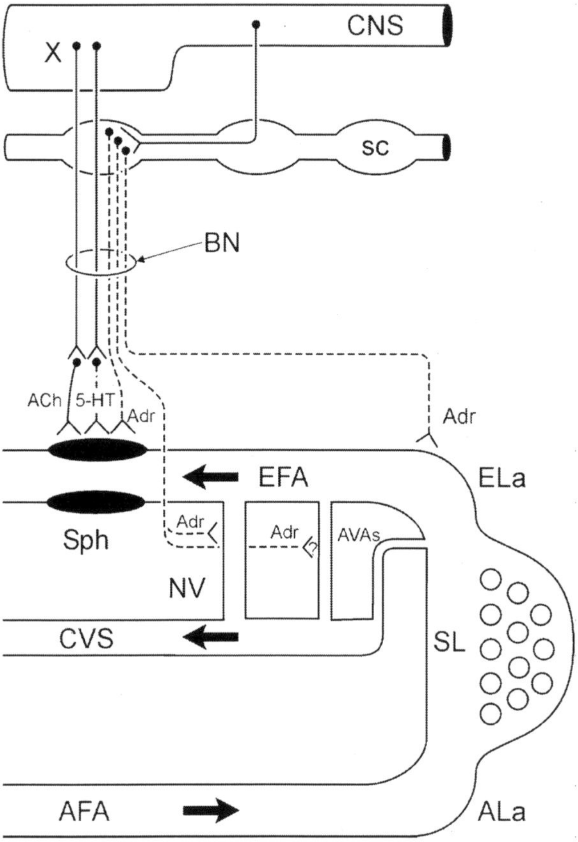

Figure: Working model for the autonomic control of teleost gill vasculature. The sphincter (Sph) at the base of the efferent filamental arteries (EFA) is the site of cholinergic (ACh), adrenergic (Adr), and serotonergic (5-HT) innervation. ACh and 5-HT innervation provide constriction, whereas Adr innervation may cause dilation, via β-adrenoceptors. Other potential sites for perfusion control are the nutritive vasculature (NV) and arteriovenous anastomoses (AVAs). Adr innervation of NV, and possibly AVAs, produces vasoconstriction via α-adrenoceptors. SLA, secondary lamella (=lamella); BN, branchial nerve; CNS, central nervous system; sc, sympathetic chain, X, vagus nerve.

Nitrergic Neurons and Neuroepithelial Cells

Branchial NANC innervation may also include nitrergic fibres, because NADPH-diaphorase reactivity was seen in 55–85% of branchial nerves in the cod, as well as in the putative sympathetic trunk (which supplies the gill arches) in stage 13 (4.5 days of development) of the spotted tilapia (*Tilapia mariae*). NADPH-diaphorase-reactive nerves also have been described in the vasculature in the gills of the Indian catfish. It has been suggested that NADPH-diaphorase histochemistry may not be as specific for nitric oxide synthase (NOS) as formerly assumed, but immunoreactive nitrergic fibres have now been described associated with the efferent filamental arteries in the Indian catfish. In addition, neuronal NOS (nNOS) immunoreactivity is seen in neuroepithelial cells adjacent to the efferent filamental artery along the length of the filament in the Indian catfish, in the interlamellar region in both the killifish and longhorn sculpin (Hyndman and Evans, unpublished data), and in cells distinct from but adjacent to Na^+-K^+-ATPase-containing MRCs in the operculum of the killifish. Partial clones for nNOS from the gills of the spiny dogfish (GenBank accession no. 232227) and killifish (GenBank accession no. AY533030) now have been reported.

There is clear evidence that either a NO donor (sodium nitroprusside, SNP) or NO itself is vasodilatory in systemic vessels (and the ventral aorta) in the rainbow trout and eel, and infusion of NOS inhibitors N^G-nitro-L-arginine methyl ester (L-NAME) or N^G-nitro-L-arginine (L-NNA) constricted the rainbow trout coronary system. On the other hand, SNP constricted the ventral aorta, anterior mesenteric artery, and posterior intestinal vein of the dogfish shark and the ventral aorta of the hagfish and produced constriction followed by dilation in the ventral aorta of the sea lamprey. In contrast to the finding that SNP dilated the eel ventral aorta, another study has found that SNP and SIN-1 (another NO donor) contracted the perfused gill preparation of the same species. The effect could be mimicked by the stable cGMP analogue 8-bromo-cGMP and inhibited by the addition of the soluble guanylyl cyclase (sGC) inhibitor 1H-[1,2,4]oxadiazole[4,3-a]quinoxalin-1-one (ODQ), demonstrating that intracellular cGMP was involved. To our knowledge, there are no published data suggesting that NO affects gill vasculature directly, but it likely does, given the extensive expression of nNOS in the gills. A more specific description of NO effects on gill vessels awaits the kind of video-microscopic studies that have been described for other effectors. Despite obvious vasoactivity of NO in fish vessels, it appears that a prostanoid, not NO, is the endothelium-derived relaxing factor in at least the rainbow trout and spiny dogfish, so it appears that branchial NO is likely produced by the NOS-containing neurons and neuroepithelial cells that have been described above.

Recent evidence suggests that NO may have a role in modulating salt extrusion by the gill epithelium. For example, if L-NAME was applied to the killifishopercular epithelium, the I_{sc} increased by 15%, suggesting tonic, inhibitory control by NO. In addition, inhibition of NOS by L-NAME reduced the inhibitory effect of subsequent addition of sarofotoxin S6c [endothelin (ET) agonist], suggesting that a component of the ET-mediated inhibition is secondary to stimulation of NO production.

Neuropeptides

Gastrin-like immunoreactivity has been described in neurons in the gills of the cod, and cholecystokinin (CCK)-8 and caerulein constricted both arterio-arterial and arteriovenous pathways in perfused gills in the cod. Moreover, sulfated-CCK-8 constricted arteries in the gills of the New Zealand hagfish (*Eptatretus cirrhatus*), suggesting an ancient origin of this putative modulatory system.

Neuropeptide Y (NPY)-like immunoreactivity has been demonstrated in the gill vessels of three elasmobranchs, and it produces concentration-dependent contraction of the afferent branchial artery of the smallspotted dogfish shark, mediated by a Y1-type receptor. To our knowledge, no data have been published suggesting a role for NPY in branchial ionic transport, but NPY does inhibit Cl^- extrusion by the spiny dogfish rectal gland, a tissue that expresses the same NaCl extrusion proteins as does the teleost gill.

Vasoactive intestinal polypeptide (VIP) immunoreactive fibres have been described in the gills of the cod, and VIP produced a concentration-dependent dilation in perfused gills of the brown trout. Interestingly, the dilation was inhibited by indomethacin, suggesting a role for prostanoids in the response. VIP (10 μM) stimulated the I_{sc} across the opercular epithelium of tilapia, but a concentration dependence was not examined. However, this finding is corroborated by the fact that VIP plays a major role in stimulation of NaCl secretion by the rectal gland of the spiny dogfish.

Adenosine

The paracrine adenosine also may affect gill blood flow. Adenosine increased branchial vascular resistance in rainbow trout, and this was associated with a vasoconstriction of the efferent filamental artery and shunting of blood to the ILV and more proximal parts of the filament, which was mediated by an A_1-type receptor. Activation of A_1 receptors also increased gill resistance (and decreased oxygen uptake) in the Antarctic borch, but adenosine did not affect the resistance in perfused hagfish gills, although it did reduce systemic resistance. Adenosine agonist assays demonstrated the presence of both A_1 (constrictory) and A_2 (dilatory) receptors in the ventral aorta of the dogfish shark so it is possible that both receptors may be in the gill vasculature. We are unaware of any published studies on the putative effects of adenosine agonists on gill epithelial transport, as has been described for ionic transport in the mammalian renal tubules.

ET

ET is as potent in constricting fish vasculature as it is in mammals, and the effects are both systemic and branchial. An early study demonstrated that rat ET-1 increased gill resistance in the rainbow trout, and rainbow trout ET (isolated from rainbow trout kidney extracts) potently (and concentration dependently) constricted a variety of rainbow trout vascular rings, including the efferent branchial artery. Interestingly, the trout vascular preparations were more sensitive to rat ET-1 than the homologous peptide, but the maximum tensions produced by both peptides were equivalent. Heterologous ET-1 also produced an increase in branchial resistance in the cod and constricted systemic vessels in two agnathans, the spiny dogfish and the eel. Agonist sensitivity suggests that the gill ET receptor in the spiny dogfish is an ET_B type. Lodhi et al. suggested that the ET receptor in the gills of the rainbow trout was not a classical ET_A type, because of its insensitivity to BQ-123, but they did not test for an ET_B-like receptor. They localised the ET receptor to the lamellae by ^{125}I-labelled ET-1 autoradiography, which is consistent with the subsequent, video-microscopic demonstration that ET-1 constricted lamellar pillar cells in both the rainbow trout and cod, thereby shunting blood to the outer margins of the lamellae. Our recent video-microscopic studies with the longhorn sculpin have demonstrated a nearly complete shutdown of flow to the distal filament and lamellae, consistent with pillar cell contraction and possibly constriction of the afferent filamental artery or prelamellar arterioles. Moreover, we have immunolocalised ET_B-like receptors to the afferent and efferent filamental arteries in the gills of the killifish and longhorn sculpin (as well as the postlamellar, arteriovenous anastamoses), but not to the pillar cells (K. A. Hyndman and D. H. Evans, unpublished data). Thus species differences may exist in the specific distribution of ET receptors in the fish gills. The site of ET synthesis appears to be vascular endothelial and neuroepithelial cells in the filamental epithelium in a variety offish species including teleosts and elasmobranchs. In Zaccone's studies, big ET (ET precursor) immunoreactivity colocalised with serotonin and neuropoeptide immunoreactivity in some neuroepithelial cells. Big ET has been localised to Na^+-K^+-ATPase-containing MRC in the gill epithelium of the Atlantic stingray.

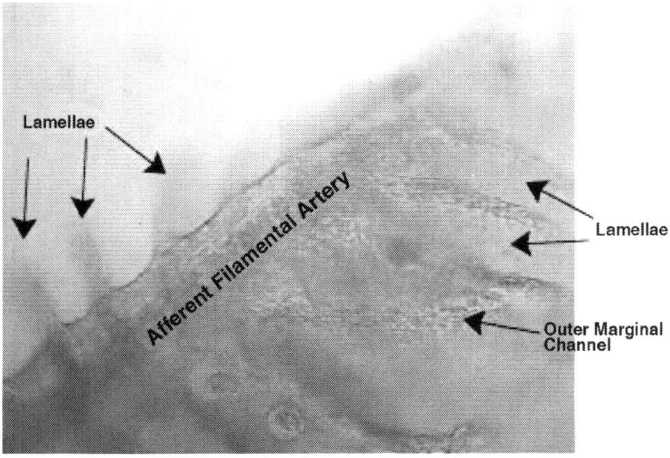

Figure: Video of the effect of endothelin on the flow of blood through the afferent filamental artery and a series of afferent lamellar arterioles leading into lamellae of a single filament in the gill of the longhorn sculpin *Myoxocephalus octodecimspinosus*.

Endothelin has direct effects on mammalian renal transport, and recent data suggest that epithelial effects of ET can be demonstrated in fishalso. Specifically, ET-1 inhibited the I_{sc} produced by the killifish opercular epithelium in a concentration-dependent manner, and the effect is mimicked by the ET_B-specific agonist sarafotoxin S6c, suggesting a role for ET_B-like receptors.

Superoxide

It is now well established that cellular production of superoxide ions (O_2^-) has direct metabolic effects and indirect effects because of the instantaneous reaction of O_2^- with NO to produce the very toxic peroxynitrite ($ONOO^-$) and also reducing the NO concentration. In addition, $ONOO^-$ itself has been found to be dilatory in the rat pulmonary artery. This is a relatively unstudied area in fishes, but a recent study demonstrated that acclimation of the Adriatic sturgeon (*Acipenser naccarii*) to seawater was accompanied by a significant increase in superoxide dismutase (SOD) activity, suggesting an increased production of O_2^- during an osmotic stress. Such a response also may have vascular effects, because it has been shown recently that pretreatment with SOD potentiated the constrictory effect of the NO donor SIN-1 on the perfused gill preparation of the rainbow trout, suggesting that either O_2^- or $ONOO^-$ may be dilatory or $ONOO^-$ production reduces NO concentrations in this system. Effects of this complex chemical interaction on gill ionic transport are also possible, because we have recently demonstrated that pretreatment of the killifish opercular epithelium with the SOD mimetic TEMPOL (4-hydroxy-2,2,6,6-tetramethylpiperidine 1-oxyl) reduced the ET-induced inhibition of the I_{sc} across the tissue by 34%. Further studies on a putative role for this system on gill perfusion and transport are warranted.

Prostanoids

Branchial production of prostanoids may play a role in gill perfusion, because a cyclooxygenase (COX) inhibitor, indomethacin, attenuated the VIP-induced dilation of perfused rainbow trout gills. However, the data on the vasoactivity of prostanoids in fish vessels are conflicting. For example, Piomelli et al. demonstrated a prostacyclin (PGI_2)-induced decline in gill vascular resistance in perfused heads of two elasmobranchs, but an increase in resistance in the same preparation in four teleosts, as well as perfused arches of two other teleosts and an elasmobranch. In all cases, adrenergic receptor inhibitors had no effect, suggesting that the effects were direct and not mediated by secondary release of epinephrine. A subsequent study showed that PGE_2 increased gill resistance in eels in vivo, and this has been corroborated by a video study that showed that infusion of $PGF_{2\alpha}$, but not PGE_2, decreased the diametre of the afferent filamental arteries in the cod. In all, it appears that the prostanoids are constrictory in the gills and may have prelamellar effects on the afferent filamental artery. Gill tissue from both the rainbow trout and European eel produces prostanoids, including PGE_1, PGE_2, $PGF_{1\alpha}$, $PGF_{2\alpha}$, and PGD_2, and in the eel (but not rainbow trout) the levels of each fell when the fishwere acclimated to seawater. Using a heterologous antibody, COX immunoreactivity has been localised to the ILV of the Atlantic stingray gills and in MRCs in the killifish gills (Piermarini and Evans, unpublished data), and we have recently cloned a COX-2-like fragment from the gill of the killifish.

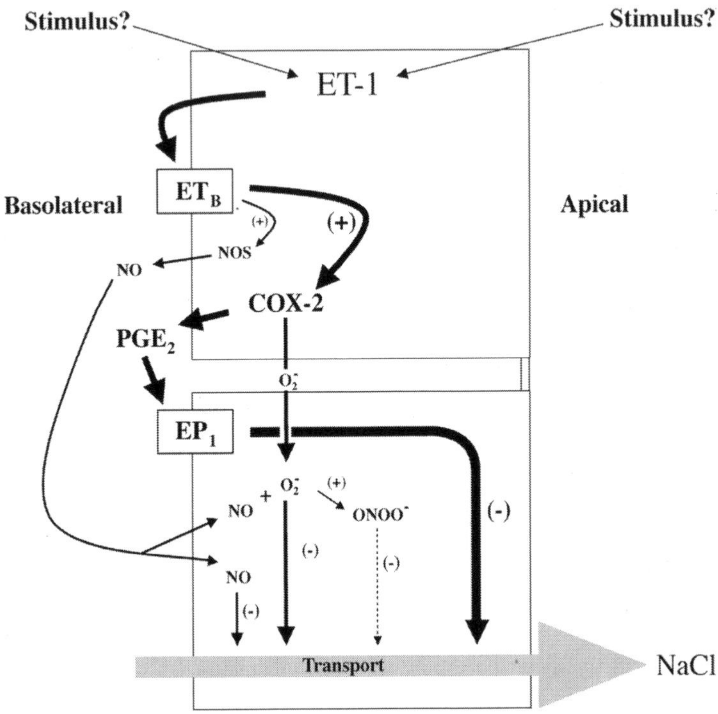

Figure: Working hypothesis for the putative pathways of endothelin (ET) inhibited NaCl transport across the fish gill. Two cells are diagrammed, but the system may be expressed within a single cell. The width of the arrows is proportional to the presumed importance of the specific pathway in the axis. ET-1, endothelin; ET_B, endothelin B receptor; NOS, nitric oxide synthase; COX-2, cyclooxygenase-2; NO, nitric oxide; PGE_2, prostaglandin E; EP_1, PGE_2 receptor; O_2^-, superoxide ion; $ONOO^-$, peroxynitriteion.

Prostanoids (specifically PGE_2) also may play an important role in modulating NaCl extrusion by the marine teleost gills. Two early studies demonstrated that PGE_2 inhibited the I_{sc} across the killifish opercular epithelium and a recent study found that PGE_2 inhibited the I_{sc} produced by a cultured epithelium of PVCs from the sea bass (*Dicentrarchus labrax*). However, it should be noted that this cultured epithelium is presumably lacking MRCs and produces a very small I_{sc} compared with the killifish opercular preparation. As mentioned above, the ET-1 (actually the ET_B agonist sarafotoxin S6c) mediated reduction of the I_{sc} across the killifish opercular epithelium can be inhibited partially by preincubation of the tissue with either L-NAME (17%) or TEMPOL (34%), suggesting that both NO and O_2^- may be secondary mediators of the ET response. Preincubation of the tissue with indomethacin inhibited the sarafotoxin S6c effect by 90%, suggesting that COX-produced prostanoids are the major effectors of this axis, just as they appear to be the dominant endothelium-derived relaxing factor (EDRF) in fishes. Moreover, PGE_2 produced a concentration-dependent inhibition of the I_{sc}. Other prostanoids were not inhibitory. The COX-2 specific inhibitor NS-398 was much more effective than the COX-1 inhibitor SC560, and an $EP_{1/3}$-receptor specific agonist, sulprostone, produced a concentration-dependent inhibition of the I_{sc}, but butaprost (an EP_2-specific agonist) produced a concentration-dependent stimulation. A working model for this new signalling axis in the fish gill is diagrammed below.

Extrinsic Control

Prolactin

We are unaware of any studies on the hemodynamic effects of prolactin in fishes, but it has been clear for 45 years that the adenohypophysial peptide plays a major role in gill function in freshwater osmoregulation. In a classic paper, Pickford and Phillips demonstrated that hypophysectomised killifishcould survive in fresh water only if they were treated with prolactin, and subsequent isotopic studies showed that this survival was associated with a reduction in ionic loss, rather than stimulation of ionic uptake. More recently, it has been shown that prolactin can also restore osmoregulatory ability to the hypophysectomised channel catfish in freshwater. Prolactin injections reduced gill Na^+-K^+-activated ATPase activity in the killifish, the thick-lipped grey mullet (*Chelon labrosus*), gilthead sea bream, and rainbow trout. However, prolactin had no effect on either expression of Na^+-K^+-ATPase (by Northern blot) or enzymatic activity in the gill epithelia of the brown trout and tilapia, as well as the opercular epithelium in tilapia. In the latter study, prolactin did not affect MRC number, but it reduced the size of individual cells, and this was corroborated by a subsequent study in a related species (*O. niloticus*) that found that prolactin injection stimulated the formation of small

MRCs, similar to the "beta cells" of freshwater acclimated teleosts. A similar cellular remodelling was produced by prolactin injection into the silver sea bream (*Sparus sarba*). These studies corroborate earlier work that found that prolactin injections into tilapia were associated with a reduction of both I_{sc} and conductance across the in vitro opercular epithelium, suggesting epithelial remodelling in addition to direct effects on ionic transport. As one might expect, plasma levels of prolactin are inversely related to salinity. A recent study has demonstrated the importance of prolactin in freshwater ion regulation directly, by showing that a cultured gill epithelium (that contains both PVCs and MRCs) can only transport Na^+ and Cl^- inward when prolactin and cortisol are in the culture medium. Interestingly, Na^+-K^+-activated ATPase activity was unaffected by these treatments.

Prolactin receptors have been cloned in several fish species, including two species of tilapia and rainbow trout, goldfish, and gilthead sea bream, and expression was high in the gill in all cases. Using either in situ hybridisation or immunohistochemistry, the receptors could be localised to the MRCs in most of these species. Interestingly, in the tilapias and rainbow trout, transfer to hyperosmotic salinities was associated with equivalent or increased expression of the receptor, suggesting that prolactin might have some role in acclimation to salinities other than fresh water.

Cortisol

Cortisol is the major corticosteroid produced by the interrenal gland in fishes and functions as both a glucocorticoid and mineralocorticoid. Until recently, cortisol has been considered to be involved primarily in seawater osmoregulation because it reversed the reduction in Na^+-K^+-activated ATPase activity produced by hypophysectomy of the killifish, stimulated the enzyme's activity in a variety of fish species, and increased tolerance to high salinities. Moreover, there was a correlation between Na^+-K^+-activated ATPase activity, Na^+ effluxes, and plasma cortisol concentration as the eel was acclimated to seawater, and there was a rapid increase in plasma cortisol concentrations preceding the increase in CFTR expression in the killifish gills after transfer to seawater, as well as associated with the changes in expression of Na^+-K^+-ATPase a_{1a}- versus a_{1b}-mRNA in the gills of the rainbow trout after transfer to seawater. Injection of cortisol stimulated Na^+-K^+-activated ATPase activity in the gills of the gilthead bream and tilapia. Cortisol also increased the expression of

NKCC (as well Na^+-K^+-activated ATPase activity) in the gills of freshwater Atlantic salmon, and the increase was associated with proliferation of MRCs in the epithelium, as had been demonstrated in other salmonids. At least some of this cytogenic effect is due to direct stimulation (rather than stimulation of other hormones in vivo), because it can be produced in the tilapia opercular epithelium cultured in vitro. These studies suggest that the effect of cortisol on the fish gills is primarily cytogenic, and this is corroborated by the finding that cortisol does not stimulate the I_{sc} produced by the killifish opreculum.

Radioligand binding protocols have identified cortisol receptors in the gills of the eel, two salmonids, and tilapia, and in tilapia, acclimation to seawater is accompanied by an increase in the number of receptors. The receptors were localised to two types of MRC (defined by Na^+-K^+-ATPase immunoreactivity) in chum salmon fry, using both immunological and molecular protocols. In this study, cortisol receptors were also localised to undifferentiated cells in the interlamellar region, suggesting a role in MRC development, which corroborated an earlier study that found that a mineralocorticoid receptor antagonist, spironolactone (but not the glucocorticoid antagonist RU486), inhibited the proliferation of MRCs in the gills of the rainbow trout that had been exposed to ion-deficient tap water. In addition to supporting a cytogenic rather than physiological role for cortisol, this study also supports the idea of a role for cortisol in freshwater fishes, which will be discussed further below. A cortisol receptor for the rainbow trout has been cloned and shares more homology with the mammalian glucocorticoid receptor (GR) than mineralocorticoid receptor (MR). However, an MR-like receptor has now been cloned from the same species, and recent studies have reported two GRs in the trout gills and four, distinct corticosteroid receptors in the cichlid *Haplochromis burtoni*, one MR and two GRs, one of which has two splice variants.

Early plasma assays found that cortisol levels increased when fish were acclimated to seawater, but more recent studies have shown that cortisol concentrations also increase after transfer to fresh water in eel, tilapia, killifish, mullet, flounder, and sea bass, suggesting a role for cortisol in freshwater osmoregulation. Early studies demonstrated that cortisol injection increased Na^+ uptake in intact and interrenalectomised European eels and intact goldfish in fresh water and cortisol treatment increased the plasma osmolarity of hypophysectomised European eels and goldfish in fresh water.

Importantly, the hypophysectomised black bullhead and Indian catfish need both prolactin and cortisol to osmoregulate in freshwater. More recent studies found that the cortisol-enhanced Na^+ uptake was associated with an increased apical MRC surface area in the rainbow trout, European eel, tilapia, and catfish, and Dang et al. demonstrated that cortisol treatment increased the intracellular tubular system in the MRCs, as well as the expression of immunoreactive Na^+-K^+-ATPase. In addition, cortisol may stimulate ionic transport steps directly, because Lin and Randall found that H^+-ATPase activity in gills homogenates was stimulated by cortisol, and Kelly and Wood found that cortisol increased Na^+ and Cl^- uptake by cultured PVCs from the rainbow trout. As noted in section VIII*C1*, this group has recently demonstrated that both cortisol and prolactin are necessary to stimulate net Na^+ and Cl^- uptake by a cultured epithelium that contains both PVCs and MRCs from trout gills.

In summary, current evidence suggests that the primary mode of action of cortisol is cytogenic, mediating the growth and differentiation of transport cells in the gill epithelium in both seawater and fresh water. In fresh water, cortisol appears to work in concert with prolactin, but in seawater another synergistic system is involved: growth hormone and insulin-like growth factor I (IGF-I).

Growth Hormone and IGF

As might be expected, fish growth hormone (GH) has morphogenic effects (independent of GH-enhanced size), and the major effect appears to be stimulation of osmoregulation in seawater. Injection of GH promotes salinity tolerance, associated with increased Na^+-K^+-activated ATPase activity and MRC size and density, in salmonids as well as two species of tilapia, the killifish, and silver sea bream. GH injections were associated with increased expression of Na^+-K^+-ATPase in MRC of the brown trout and expression of NKCC1 in the MRCs in the gills of the Atlantic salmon. GH receptors have been localised to gills in the coho salmon by radioreceptor assay, and two splice variants of a recent clone from the black sea bream (*Acanthopagrus schlegeli*) can be localised to a variety of tissues, including the gills. In the killifish, Atlantic salmon, and brown trout, some of the effects of GH can be mimicked by injection of IGF. For example, IGF-I injection stimulated MRC development and Na^+-K^+-ATPase expression in the MRC in the gills of the brown trout. Transfer to seawater is associated with an increase in plasma concentrations of GH in salmonids, as well as increased message for IGF-I in the gills of coho salmon. The gene for IGF-I and its product can be localised to MRC in the tilapia gills, and an IGF-I receptor has been cloned and localised to the gills in two sculpins. It is generally considered that IGF-I is the direct mediator of increased Na^+-K^+-ATPase activity in the MRCs but that GH is needed for the response.

The fact that cortisol and GH together stimulate salinity tolerance and Na^+-K^+-activated ATPase activity better than either one alone in various salmonids and the killifish (and NKCC1 expression in the Atlantic salmon) suggests that cortisol and GH interact to promote seawater osmoregulation. There also appears to be some synergism between cortisol and IGF-I injections. One site of GH/cortisol interaction may be cortisol receptors because GH treatment was associated with an increase in the number of cortisol receptors in the gills of the coho and Atlantic salmon. The synergism may be cytogenic, however, because GH increased general mitotic activity in the gills of the rainbow trout, but cortisol specifically stimulated the number of MRC.

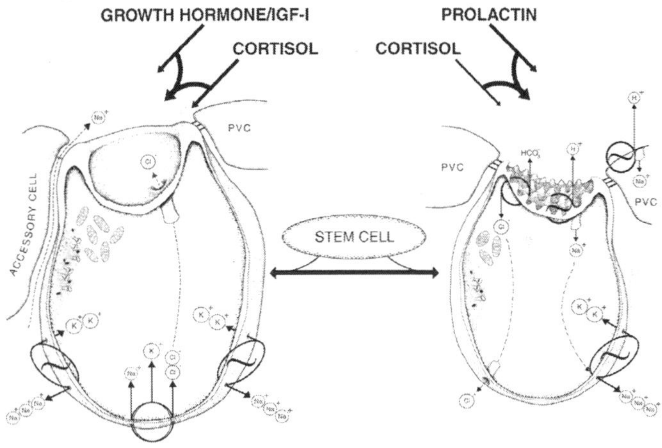

Figure: Working model for the effects of prolactin, cortisol, growth hormone, and insulin-like growth factor I on the morphology and transport mechanisms of gill mitochondrion-rich cells in the gill epithelium of teleost fish. In seawater, MRCs are generally larger and contain a deep apical crypt; in fresh water, the apical surface is broad and contains numerous microvilli. Growth hormone and cortisol can individually promote the differentiation of the seawater MRC, and also interact to control epithelial transport capacity. Prolactin inhibits the formation of seawater MRCs and promotes the development of freshwater MRCs. Cortisol also promotes acclimation to fresh water by maintaining ion transporters and MRCs, and by interacting to some degree with prolactin.

Angiotensin

Despite earlier uncertainty, it is now clear that a renin-angiotensin system (RAS) is expressed in fishes, including elasmombranchs and agnathans, although fish produce little or no aldosterone. ANG II constricted perfused gills in both the trout and eel, but local, tonic production of ANG II appears to be relatively unimportant in gill hemodynamics, because inhibition of ANG II production in the trout in vivo by the angiotensin converting enzyme (ACE) inhibitor lisinopril did not affect gill resistance, although systemic resistance was reduced by nearly 50%.

Fish ANG II has three amino acid substitutions in the basic octapeptide, compared with mammals. There generally is a direct correlation between plasma ANG II (and renin) levels and salinity in fishes, including the river lamprey, two eel species, and two salmonids. Enzymatic analyses have found ACE activity in the gills of the trout and river lamprey as well as in the corpuscles of Stannius in the rainbow trout and American eel. Fish corpuscles of Stannius are located within the renal tissue of teleosts and are the site of the production of a unique hormone, stanniocalcin, important in Ca^2 regulation by the gills. Within the rainbow trout gill, ACE can be localised to pillar cells nearest the inner lamellar margin and, to a lesser extent, the lumen of the afferent filamental artery and afferent lamellar arterioles.

ANG II receptors have been localised to gill tissue in the banded houdshark (*Triakis scyllia*) by radioligand binding, and antibodies raised against the mammalian AT_1 receptor localised the receptor to MRCs and PVCs in the gill epithelium in the eel and icefish (*Chionodraco hamatus*). A partial sequence for the eel ANG II receptor has been reported (Tran Van Chuoi et al., GenBank accession no. AJ005132), but no expression data are available. In the eel, acclimation to seawater was associated with a threefold increase in immunoreactive AT_1 in the gills. Infusion of ANG II into the European eel was associated with a concentration-dependent increase in Na^+-K^+-activated ATPase activity in the gills of freshwater acclimated fish, but a biphasic (stimulation inhibition) response with the same concentrations in the seawater acclimated fish. This may appear to be counter to the finding that ANG II inhibited the transepithelial potential across perfused flounder gills, but it is generally accepted that the transepithelial potential in vivo is the result of differential gill ionic permeabilities, not active transport. Unfortunately, no data have been published on the effect of ANG II on the I_{sc} across the killifish opercular epithelium.

Bradykinin

At least some components of the kallikrein-kinin cascade are present in the plasma of teleosts, lungfish, and lower actinopterygians such as gar, bowfin, and sturgeon, but apparently not in the plasma of either agnathans or elasmobranchs. The sequences of trout, cod, and lungfish bradykinin (BK) are now published, and it appears that trout may express two kinogen genes, which produce distinctive BK precursors. Infusion of homologous BK into the trout produced a triphasic response (pressor-depressor-pressor) in dorsal aortic pressure. The second phase was characterised by a slight increase in gill resistance, which was abolished by pretreatment with indomethacin, suggesting a role for prostanoids in the response. Isolated, efferent branchial arteries from the trout were unresponsive to homologous BK, and we found that aortic rings from the spiny dogfish were unresponsive to mammalian BK (0.1 nM to 0.1 μM; D. H. Evans and J. Nowacki, unpublished data). Thus a role for BK in modulating gill perfusion remains unclear, despite the fact that there is clear activity in other vascular beds. Structure-activity protocols suggest that the fish BK receptor is different from either B_1 or B_2 receptors, but a B_2 ortholog has been sequenced (and functionally expressed) recently from the zebrafish. There are no published data suggesting that BK may modulate salt transport across the fish gills, but it has been shown to activate a Ca^{2+}-dependent Cl^- conductance in the mouse inner collecting duct epithelium, so further investigation is warranted.

Arginine Vasotocin

Like all nonmammalian vertebrates, fishes (including the agnatha) produce the neurohypophysial arginine vasotocin (AVT), with a single amino acid substitution compared with vasopressin. AVT is one of the most potent vasoconstrictors in the teleost gills, decreasing flow through the arterio-arterial pathway in the eel gills in vivo and decreasing flow through the arteriovenous pathway in the rainbow trout in vivo at low doses and through the arterio-arterial pathway at higher concentrations. In the trout, a bolus injection of AVT doubled the total gill resistance, while only increasing systemic resistance by 15%. AVT also contracted ventral aortic and efferent branchial arterial rings from the trout. The pharmacological profile of the effect of AVT on the trout efferent branchial artery suggested a receptor similar to the mammalian arginine vasopressin (AVP) V1a and oxytocin receptors, but an earlier study of the effects of putative AVP receptor agonists and antagonists on adenylate cyclase activity in rainbow trout gill plasma membranes suggested that the receptor was distinct from either V1 or V2 receptors. However, a V1-type receptor has now been cloned from the European flounder, and its message could be localised to the gills. The relationship between fish plasma AVT levels and external salinity is unclear. Early radioimmunoassay studies, using a fish-specific antibody, found no difference in plasma AVT concentrations in seawater versus freshwater acclimated European flounder, rainbow trout, and European eel, and an ELISA protocol confirmed this finding for the rainbow trout. In contrast, other studies determined that plasma AVT concentrations increased when the European flounder or rainbow trout was acclimated to fresh water, and a recent study showed a direct correlation between plasma AVT levels and salinity in the European flounder. Some of these discrepancies may be due to diurnal changes in plasma levels. Recent evidence suggests that AVT also may play a role in modulating salt transport across the fish gill epithelium, although the current model system consists of a cultured epithelium of PVCs from the sea bass, which maintains an extremely low I_{sc} compared with the killifish operculum (3–5 vs. >50 μA/cm^2). In the sea bass preparation, AVT stimulates the I_{sc} and the agonist profile suggests mediation by a V1-type receptor. Effective doses, however, were much higher than those measured in plasma.

Natriuretic Peptides

Various components of the natriuretic peptide (NP) system are expressed in all fish groups that have been examined to date (including the elasmobranchs and agnatha), and NPs play important roles in cardiovascular, renal, and gill physiology. Amino acid sequences have been published for atrial NP (ANP) in the Japanese eel (*Anguilla japonica*) and rainbow trout, and for C-type NP (CNP) in four species of elasmobranchs and three species of teleosts. Four distinct CNPs have been cloned recently from the medaka (*Oryzias latipes*) and fugu. Brain NP (BNP) has not been isolated from fish, but a unique NP, termed VNP, has been isolated from ventricular tissue from the Japanese eel and two species of salmonids. Teleosts produce both ANP and CNP, but elasmobranchs apparently express only CNP, and a recent study suggests that the agnathan NP is unique. Four NP receptors have been cloned from the Japanese eel (termed NPR-A through D), with mammalian-like structures and ligand specificities in the first three. NPR-D has only been described in fishes and is closely related to NPR-C, sharing a short cytoplasmic carboxy-terminal tail that does not include the guanylyl cyclase catalytic domain. NPR-B has been cloned from the rectal gland of the spiny dogfish shark and from the medaka, and partial sequences for an NPR-A/B and NPR-C/D have been reported from the New Zealand hagfish (*Eptatretus cirrhatus*) and pouched lamprey, respectively. A partial sequence for a putative NPR-C/D also has been reported for the spiny dogfish. Radioligand binding protocols have localised NPR-A/B and NPR-C / D receptors in the gills of gulf toadfish, Atlantic hagfish, pouched lamprey, and New Zealand hagfish. It is generally assumed that the CNP/NPR-B couple is the ancestral vertebrate condition.

It has been pointed out that the use of heterologous antibodies for radioimmunoassay determination of plasma NP concentrations often has produced conflicting results, and recent results, using homologous antibodies for Japanese eel peptides, have shown that salinity changes in this species are associated with differential secretion of specific NPs. For instance, the hypernatremia associated with acclimation to seawater is associated with an increase in ANP secretion (and a secondary stimulation of cortisol secretion); acclimation of eels to freshwater is associated with an upregulation of both CNP expression and its specific receptor, NPR-B. NPs are potent dilators of vascular smooth muscle in fishes, including teleosts, elasmobranchs, and agnathans, and ANP decreased gill resistance in the rainbow trout in vivo, corroborating an earlier study that demonstrated that ANP decreased arterio-arterial resistance in the epinephrine-contracted, perfused gills from that species. Unfortunately, no in vivo videomicroscopy studies have been done to visualise the effects of NPs on gill blood flow directly.

Since an early study found that a truncated ANP stimulated the I_{sc} across the killifish opercular epithelium it has been accepted generally that one of the prime roles of NPs in seawater teleosts is the stimulation of gill Na^+ excretion by the branchial epithelium. However, a recent study was unable to elicit any effect on the I_{sc} across this epithelium by either eel ANP or porcine CNP (0.1 nM to 0.2 μM), so it is not clear that NPs stimulate salt extrusion by the marine teleost gills, despite the fact that ANP apparently may increase the Na^+-K^+-activated ATPase activity in isolated gill cells from the eel (Flik and Takei).

Thyroid Hormones

No data have been published suggesting that thyroid hormones affect gill perfusion, but they may play an indirect, stimulatory role in gill salt extrusion. For instance, thyroxine (T_4) or triiodothyronine (T_3) treatment stimulated Na^+-K^+-activated ATPase activity and/or MRC number in the Atlantic salmon, rainbow trout, climbing perch, and tilapia , as well as immunoreactive Na^+-K^+-ATPase in MRC in the gills of the summer flounder (*Paralichthys dentatus*). However, it appears that these effects may be mediated via the cortisol and/ or GH-IGF axes, although at least one study found that the effects on Na^+-K^+-activated ATPase activity of GH and T_3 were additive. For instance, T_4 treatment alone did not stimulate Na^+-K^+-activated ATPase activity in the gills of tilapia, but it potentiated the stimulation produced by cortisol, as it did the effect of GH on enzyme activity in the amago salmon (*Oncorhynchus rhodurus*). Moreover, T_3 treatment increased the number of cortisol receptors in the gills of the Atlantic salmon, which was further potentiated with GH addition, and it increased the number of cortisol receptors (and the cortisol stimulation of Na^+-K^+-activated ATPase activity) in rainbow trout gills.

Glucagon

As is true for thyroid hormones, pancreatic glucagon has not been shown to affect gill perfusion, but this peptide may play a role in gill transport. In an early study of the tilapia opercular epithelium, Foskett et al. found that glucagon could stimulate the I_{sc} across the epithelium in a concentration-dependent manner, with a minimum effective dose of 1 nM and a maximum stimulation of 72% at 10 μM. Moreover, glucagon stimulated adenylate cyclase in the rainbow trout gills, as it does in the process of stimulating NHE3 in the mammalian kidney, a transporter of relevance to fish gills.

Urotensins

Since the initial characterisation of urotensins (UT) from the fish urophysis, it has become apparent that these neuropeptides may be major cardiovascular effectors in fishes and amphibians, as well as mammals. In fishes, urotensins (UI and UII) are secreted by neurosecretory cells (Dahlgren cells) in the distal spinal cord, which secrete into the neurohemal urophysis in most groups. In amphibians and mammals, U1 and related peptides (CRF, sauvagine, urocortin) are expressed in the central nervous system, and this apparently is also true in both teleosts and elasmobranchs. UT-secreting neurons have not been described in fish gills to date. UI and/or UII have now been sequenced from many fish species, including teleosts, elasmobranchs, and lampreys, and UI-related receptors (CRF-like receptors) have been cloned from the chum salmon and brown bullhead catfish. Infused UII (a somatostatin-like peptide) produced an increase in gill vascular resistance in the rainbow trout in vivo and also a concentration-dependent contraction of isolated, efferent branchial artery rings, and UII contracted rings from the afferent branchial artery from the smallspotted dogfish shark. These data are consistent with the recent demonstration that UII is a very potent vasoconstrictor in mammals. However, some of the vascular effects of UII may be secondary to stimulation of the parasympathetic nervous system, because á-adrenergic blockade completely abolished the UII generated hypertension in the smallspotted dogfish shark in vivo. UI may be a vasodilator, because infusion into the trout was associated with vascular hypertension, which became hypotension when an á-adrenergic blocker was applied. This study corroborated earlier data from the smallspotted dogfishshark, which demonstrated a transient hypotension, followed by an á-adrenergic-mediated hypertension, subsequent to infusion of UI. None of these experiments examined gill responses specifically, but they suggest that the two UTs may have opposing vascular effects.

The fact that immunoreactive UII appeared to be greater in the urophysis of seawater species versus freshwater species suggested an osmoregulatory role for at least UII. This was corroborated by a subsequent study that demonstrated increased immunoreactivity for both UI and UII in the urophysis of freshwater acclimated longjaw mudsucker, but this could be secondary to increased synthesis and storage, decreased release of stored peptide, or decreased peptide degradation. Similar results were published for the euryhaline bogue (*Boops boops*). Subsequent determinations of plasma UT concentrations by homologous radioimmunoassay in the euryhaline, European flounder have found a direct correlation between plasma osmolarity and UII levels. UII inhibits the I_{sc} across the seawater longjaw mudsucker jaw skin epithelium, and UI stimulates the I_{sc} but both UI and UII stimulate cortisol secretion by the rainbow trout or European flounder interrenal gland, suggesting that UTs may have both direct and indirect effects on fish osmoregulation. Because the urophyses from brook trout (*Salvelinus fontinalis*) from acidified lakes in Canada displayed lower UII content, it may be that UTs also play a role in acid-base balance. Further studies clearly are warranted.

Calcitonin and Calcitonin Gene-related Peptide

Calcitonin (CT) is produced in the ultimobranchial glands of fishes and tetrapods, except for mammals, where it is synthesized in the parafollicular "C" cells in the thyroid gland. Agnathans lack the gland, and they were thought to lack CT until it was recently immunolocated in the plasma of freshwater (but not seawater)-acclimated, Japanese lamprey (*Lampetra japonica*) and also in the inshore hagfish (*Eptatretus burgeri*), where concentration was correlated with plasma Ca^{2+} levels. Calcitonin immunoreactivity has been measured in a number of species of teleosts and elasmobranchs and the peptide sequence has been determined for a number of species, including fugu. The CT gene also has been sequenced from many species, and the gene and its transcript can be localised to the gills of the Atlantic salmon, so the ultimobranchial gland may not be the only site of CT production. Moreover, a CT-receptor like gene has been cloned from the gill cDNA library of the Japanese flounder (*Paralichthys olivaceus*).

Calcitonin is hypocalcemic in tetrapods, and current evidence suggests that this also is the case in fishes, although hypercalcemic effects are sometimes described. Infusion of Ca^{2+} into the Japanese eel was associated with an increase in plasma CT immunoreactivity, but transfer of this species from freshwater to seawater did not elicit a change in plasma CT levels, despite increased plasma Ca^{2+} concentrations. However, transfer of rainbow trout to seawater was associated with significant increases in plasma CT concentrations. Several studies have demonstrated that CT inhibits Ca^{2+} uptake by the gills, consistent with a hypocalcemic role.

Calcitonin gene-related peptide (CGRP), a splice variant of the CT mRNA, is also found in fishes and may play a role in vascular dynamics, osmoregulation, Ca^{2+} balance, and gas exchange. CGRP-containing fibres have been described in the brain and/or gut of all the major fish taxa. The peptide also has been identified in the plasma of the pink salmon (*Oncorhynchus gorbuscha*) and in gills of the rainbow trout by radioimmunoassay, where it binds specifically and stimulates adenyl cyclase by a GTP-binding process. The gene for a CGRP receptor has been cloned from the gills of the Japanese flounder. Acclimation of the rainbow trout to seawater is associated with an increase in plasma levels of CGRP as well as specific binding to gill membrane receptors. Interestingly, this study also found increased gill CA levels in seawater, and this group previously had demonstrated that CGRP can stimulate production of this enzyme in trout gill extracts. Finally, it is possible that CGRP could be vasodilatory in the fishgills, as it is in small arteries from the rainbow trout gut.

Stanniocalcin

Stanniocalcin (STC, formerly termed hypocalcin or teleocalcin) is a glycoprotein that is produced by the corpuscles of Stannius, which are located in renal tissue in teleost and holostean (gar, bowfin) fishes only. However, it is now clear that STC and its receptors also are expressed in a variety of mammalian (and human) tissues, including kidney and gut, and play roles in mineral metabolism, neural differentiation, reproduction, and even cancer.5 It has been demonstrated that CS extracts inhibit Ca^{2+} uptake by three salmonids and the American eel, and stanniectomy of the eel was associated with increased plasma Ca^{2+} levels, which were corrected when STC was injected, so STC is generally considered to be the major hypocalcemic factor in fishes. The cellular site of STC production in the CS has been localised by immunocytochemistry, and hypercalcemia in the rainbow trout is associated with increased release of immunoreactive STC from the CS. Moreover, intraperitoneal injection of $CaCl_2$ into the rainbow trout and European eel stimulated STC release from the CS, as did acclimation of the eel to seawater versus fresh water. It appears that the Ca^{2+} stimulation of CS secretion of STC may be direct and/ or via cholinergic innervation, but it also may involve direct stimulation of transcription. A calcium-sensitive receptor (CaR) has been sequenced from rainbow trout CS, and transcripts are also expressed in the gills. A CaR was concurrently cloned from dogfish shark, winter flounder, and Atlantic salmon and expression shown in the kidney and gills, which was suggested to play a role in salinity perception. Corpuscles of Stannius extracts also produce hypocalcemia in the Japanese stingray (*Dasyatis akajei*;). The gene for coho and chum salmon STC has been sequenced, and initial studies on the distribution of the STC mRNA and product suggested that they are expressed only in cells in the CS in fishes. However, more recent in situ hybridisation and Northern blot studies have delineated STC expression in a variety of tissues in the rainbow trout and arawana (*Osteoglossum bicirrhosum*), including the kidney and gills, suggesting that STC may act as a paracrine as well as circulating hormone. Interestingly, two distinct STCs are expressed in the CS of the Atlantic salmon. STC is thought to inhibit Ca^{2+} uptake across the apical membrane of the MRC, because stanniectomy in the European eel is not associated with any measurable changes in either Ca^{2+}-activated ATPase or Na^+/Ca^{2+} exchanger, both of which are thought to be basolateral. These authors conclude that the increased Ca^{2+} influx subsequent to stanniectomy is secondary to an increase in the number and/or size of the MRCs.

Parathyroid Hormone-related Protein

Fishes lack an encapsulated parathyroid gland, but parathyroid hormone (PTH)-like immunoreactivity has been reported in the pituitaries and plasma of several teleosts. However, the recently described PTH-related protein (PTHrP) is also a hypercalcemic factor in humans, shares some common regions at the amino terminus, and binds to a common PTH/PTHrP receptor. PTHrP-like immunoreactivity has now been localised in the pituitary, plasma, or other tissues in a variety of fishes, including teleosts, elasmobranchs, and agnathans, and immunoreactivity was localised to the gill epithelium in some of these studies, with expression confirmed by molecular techniques. The gene for PTHrP was initially cloned from fugu, but a sequence now has been reported from the gilthead sea bream, and in situ hybridisation has localised expression to, among other sites, MRCs in the gill epithelium. A radioimmunoassay also has been developed and validated for the peptide transcribed from this sequence, and immunoreactivity was localised to the gills, as well as the pituitary, esophagus, kidney, and intestine. Two relevant receptors have now been cloned from the zebrafish, one more sensitive to fugu PTHrP than human PTH and another sensitive to human PTH and not human PTHrP.

In an unpublished study, it was shown that PTHrP-like immunoreactivity in the plasma of the European flounder was inversely correlated with salinity (highest in gill tissue from freshwater acclimated individuals), suggesting that the peptide may play some role in Ca^{2+} regulation, which differs with salinity. Exposure of larval gilthead sea bream to an amino-terminal peptide from the fugu sequence increased $^{45}Ca^{2+}$ influx in both 100 and 33% seawater-acclimated fish and reduced

Ca^{2+} efflux from the seawater-acclimated fish, both in a concentration-dependent manner. To our knowledge, these are the only published data demonstrating that PTHrP is a functional hypercalcemic factor in fishes.

Other Hypercalcemic Factors

In artificial, calcium-deficient seawater, hypophysectomised killifish displayed reduced plasma Ca^{2+} levels, without any discernable changes in other plasma electrolytes, and the hypocalcemia was not seen in seawater with normal Ca^{2+}levels or after replacement therapy with either pituitary homogenates or heterologous prolactin. Ovine prolactin injections were associated with hypercalcemia in the intact European eel, stimulation of Ca^{2+} uptake from the fresh water, and increased branchial Ca^{2+}-activated ATPase activity. Similar results have been published for tilapia and carp. As one might expect for a putative hypercalcemic factor, pituitary prolactin content is often inversely related to external Ca^{2+} concentrations.

Somatolactin (SL) is a novel pituitary protein that was initially isolated from the pars intermedia of the cod and has been described in the pituitary of a variety of fish species. It is structurally related to both prolactin and growth hormone and was initially described only in teleosts, but has more recently been found in the African lungfish pituitary, suggesting that it may be described in tetrapods in the future. The fact that low environmental Ca^{2+} levels are associated with activation of SL cells, and increased mRNA levels, in the pituitary of the rainbow trout (as well as plasma SL levels) suggests that SLs is a hypercalcemic factor. Somatolactin also may function in reproduction, acid-base regulation, and stress responses in fishes.

Cortisol also may be a hypercalcemic factor in fishes, because plasma concentrations of cortisol increased when the rainbow trout was acclimated to freshwater with a reduced Ca^{2+} concentration, and 8 days of this treatment resulted in a stimulation of branchial Ca^{2+} transport and Ca^{2+}-activated ATPase activity. In freshwater with normal Ca^{2+} concentrations, treatment with cortisol was associated with a stimulation of Ca^{2+} uptake, branchial transport capacity, and hypercalcemia. Although these effects may be related to the cytogenic effects of cortisol, it has been shown recently that cortisol treatment is rapidly (30 min.) followed by a stimulation of gill Ca^{2+}-activated ATPase activity, suggesting that the cortisol effect is not via changes in transcription.

Metabolism of Intrinsic Signalling Agents and Xenobiotics

Largely due to studies from a single research group, it has become clear that the fish gill vasculature also is the site of substantial metabolism and/or inactivation of circulating and paracrine signalling agents. As we have stated previously, the gill vasculature receives 100% of the cardiac output, and it can be calculated that the vascular surface area of a 300 g rainbow trout may be 900 cm^2, nearly twice the respiratory surface area. Using perfused gill arches or cannulated free-swimming rainbow trout, the Olson group has measured fractional extractions of a variety of molecules of physiological importance. For instance, 35% of infused ANG I is converted to ANG II in the arterio-arterial pathway during a single pass through the perfused rainbow trout gills, and 60% of infused ANP, 55% of infused ET-1, and 80% of infused serotonin is extracted. The extraction of other peptides, amines, purines, and COX products is somewhat less. The ANG conversion is via ACE, which can be localised to the lamellar pillar cells and NPR-C/D type receptors in the gill vasculature presumably function as clearance receptors as they do in mammalian vasculature. It is interesting to note that branchial clearance of plasma ANP and VNP can be demonstrated in freshwater Japanese eels but not in eels acclimated to seawater, suggesting that there may be physiological control of gill clearance of at least NPs. The ET presumably is extracted by the very abundant ET_B receptors that can be localised to both afferent and efferent filamental arteries.

The fish gill is also a major site of the damage produced by environmental pollutants, and recent evidence suggests that it may be a site for metabolism and/or excretion of toxins. For instance, cytochrome *P*-450 enzymes (e.g., CYP1A) have been localised in the pillar cells and vascular endothelium of the rainbow trout and scup (*Stenotomus chrysops*), and β-naphthoflavone induced a 10-fold increase in the enzyme. In a subsequent study of the scup, gill CYP1A produced epoxyeicosatrienoic acids (EETs), which have been shown to affect renal transport and hemodynamics. Other environmental contaminants, such as 2,3,7,8-tetrachlorodibenzo-*p*-dioxin (TCDD) or polychlorinated biphenyl (PVCB) can induce CYP1A expression in the zebrafish, tilapia, and rainbow trout gills. Sadly, CYP1A has been found in gill tissue from oceanic, mesopelagic fishes of the western North Atlantic, suggesting a possible global occurrence of the contamination problem. Metallothioinein (MT), the heavy metal binding protein, can be localised to MRCs in the Atlantic salmon, and treatment with cadmium was associated with increased MT expression. Treatment of tilapia with copper induced MRC MT expression, as well as increased turnover of MRCs. Members of the multidrug resistant protein family of toxin export carriers (e.g., P-glycoprotein and Mrp2) have not been identified in the fish gills to date (D. Miller, personal communication), despite their expression in the fish brain and liver. It is clear that more investigations of the roles of the fish gill vasculature and epithelium in the metabolism of intrinsic signalling agents and xenobiotics are needed.

The fish gill is a morphologically and functionally complex tissue that is the site of numerous, interconnected physiological processes, which are vital to maintaining systemic homeostasis in the face of changing internal (e.g., acidosis) and environmental (e.g., salinity) conditions. Early studies established a basic understanding of these processes, and during the past half-century, with the development of cellular, biochemical, and molecular techniques, our understanding of the specific pathways and control mechanisms involved has been greatly enhanced. Many of these mechanisms are homologous to those found in mammals (and other tetrapods), but important differences do exist, largely due to an aquatic versus terrestrial evolution and existence. Thus the general morphometric and irrigation/perfusion constraints of gas exchange are mediated by respiratory structures and processes similar to those in mammals, but the structures and processes in fish are adapted to the specifics of aquatic versus aerial gas exchange and mediated by gills rather than by lungs.

The features of the gills that enhance gas exchange also make the gills susceptible to osmotic and ionic movements between the environment and extracellular fluids; thus osmoregulation is necessary. The branchial epithelium expresses an array of transporters and channels that are homologous to those in the mammalian kidney and balance the passive movements of water and solutes across the tissue (in coordination with intestinal and renal processes to maintain blood tonicity). The gill epithelium is also the primary site of plasma pH regulation and excretion of excess nitrogen, using transporters and channels that probably arose in marine ancestors, but have been secondarily adapted for osmoregulation in fresh water species.

These numerous functions by the gills require control and coordination by a complex web of neural, endocrine, and paracrine signalling pathways that are expressed in functionally analogous systems in mammals. However, some control systems (e.g., urotensins and stanniocalcin) were first described in fishes and are now known to play pivotal roles in mammalian physiology. Thus physiological research on "lower" vertebrates, such as fishes, is sure to lead to important discoveries in the realm of mammalian physiology.

Stress Response in Fish

The stress response in teleost fish shows many similarities to that of the terrestrial vertebrates. These concern the principal messengers of the brain-sympathetic-chromaffin cell axis (equivalent of the brain-sympathetic-adrenal medulla axis) and the brain-pituitary-interrenal axis (equivalent of the brain-pituitary-adrenal axis), as well as their functions, involving stimulation of oxygen uptake and transfer, mobilisation of energy substrates, reallocation of energy away from growth and reproduction, and mainly suppressive effects on immune functions. There is also growing evidence for intensive interaction between the neuroendocrine system and the immune system in fish. Conspicuous differences, however, are present, and these are primarily related to the aquatic environment of fishes. Toxic stressors are part of the stress literature in fish more so than in mammals. This is mainly related to the fact that fish are exposed to aquatic pollutants via the extensive and delicate respiratory surface of the gills and, in seawater, also via drinking. The high bioavailability of many chemicals in water is an additional factor. Together with the variety of highly sensitive perceptive mechanisms in the integument, this may explain why so many pollutants evoke an integrated stress response in fish in addition to their toxic effects at the cell and tissue levels. Exposure to chemicals may also directly compromise the stress response by interfering with specific neuroendocrine control mechanisms. Because hydromineral disturbance is inherent to stress in fish, external factors such as water pH, mineral composition, and ionic calcium levels have a significant impact on stressor intensity. Although the species studied comprise a small and nonrepresentative sample of the almost 20,000 known teleost species, there are many indications that the stress response is variable and flexible in fish, in line with the great diversity of adaptations that enable these animals to live in a large variety of aquatic habitats.

Bibliography

Anderson, A. Trevor: *Fish Nutrition in Aquaculture*, Springer, Delhi, 2009.

Bailey, MT: *A-Z of Tropical Fish Diseases and Health Problems*, Howell Book House, New York, 2001.

Balfour, F. M.: *A Monograph on the Development of Elasmobranch Fishes*: Macmillan, London, 1878.

Barlow, S.M.L: *Chemical and Biological Assay Procedures for Lysine in Fish Meals*, J.Sci.Food Agric., 1984.

Basant, Puri K.: *The Natural Way to Beat Depression: the groundbreaking discovery of EPA to change your Life,* Hodder and Stoughton, London, 2004.

Boserup, E.: *The Conditions of Agricultural Growth: The Economics of Agrarian Change under Population Pressure*, Aldine Press, Chicago, 1965.

Bowman, S.J.: *Digest of Statistics of IAFMM*, Potters Bar, U.K., IAFMM, 1984.

Bruno D.W., Poppe T.T., *A Color Atlas of Salmonid Diseases*, Academic Press, London, 1996.

Cech, JJ and Moyle, PB: *Fishes, An Introduction to Ichthyology*, Benjamin Cummings, NY, 2003.

Chow, Ching Kuang: *Fatty Acids in Foods and Their Health Implications*, Routledge Publishing, New York, 2001.

Clover, Charles: *The End of the Line: How Overfishing is Changing the World and What we Eat*, Ebury Press, London 2004.

Ferguson, H.W.: *Systemic Pathology of Fish*, Iowa State Press, Ames, Iowa, 1989.

Geran, James: *The Proper Care of Goldfish*, TFH Publications, NJ, 2000.

Hamm, Wolf: *Trans Fatty Acids*, Blackwell, NY, 2008.

Haskell, B.D.: *Ecosystem Health: New Goals for Environmental Management*, Island Press, Washington, D.C., 1992.

Hasler, A.D.: *Orientation and Fish Migration*, Academic Press, New York, 1971.

Kulkarni, G.K.: *Fisheries and Fish Toxicology*, APH, Delhi, 2006.

Lloyd, G.: *Mammals of the Pacific States,* Stanford University Press. 1965

Maddy Hargrove: *The Discus: An Owner's Guide to Happy Healthy Fish*, Howell Books, NY, 1999.

Mahadevan, R.K. and Selvamani, B.R.: *Fish and Fishery Culture: Assessment and Evaluation*, Campus Books International, Delhi, 2008.

Malvee, Sangeeta: *Fish Genetics*, SBS Pub, Delhi, 2008.

Martin, A.M.: *Fisheries Processing: Biotechnological Applications*, Chapman and Hall, Delhi, 2009.

Mathur, A.K. and Sharma, L.L.: *Hand Book of Freshwater Ornamental Fishes,* Yash Pub house, Delhi, 2006.

Mehta, Varun: *Fisheries and Aquaculture Biotechnology*, Campus Books International, Delhi, 2006.

Mori, Fumitashi and Sakamoto, Yohei: *Aquarium Fish of the World: The Comprehensive Guide to 650 Species*, Chronicle Books, UK, 1991.

Moyle, PB and Cech, JJ: *Fishes, An Introduction to Ichthyology*, Benjamin Cummings, NY, 2003.

Nelson, J. S.: *Fishes of the World*, John Wiley & Sons, Inc.. , NY, 2006.

Petr, Tomi: *Fisheries in Irrigation Systems of Arid Asia*, Daya, Delhi, 2007.

Pollard, D. A. *Freshwater Fishes of South-eastern Australia*. McDowell, R. M. Reed, Sydney. 1980.

Ponniah, A.G.: *Fish Biodiversity of India*, NBFGR, 2002.

Ralph, R.: *Breeding Australia's desert goby*, Tropical Fish Hobbyist, 1988.

Rao, D.V.: *Poisonous and Venomous Fishes of Andaman Islands, Bay of Bengal,* Zoological Survey of India, 2003.

Rebelin, W.E.: *The Pathology of Fishes*, The University of Wisconsin Press, UK, 1975.

Rosen, D. E.: *Modes of Reproduction in Fishes, How Fishes Breed*, Natural History Press, New York, 1966.

Sharma, Arti: *Fishes: Aid to Collection, Preservation and Identification*, Daya, Delhi, 2006.

Singh, Braj Prasad: *Fish Genetic and Biotechnology*, Swastik, Delhi, 2008.

Stoskopf, M.K.: *Fish Medicine*, W.B. Saunders Co. 1993.

Tunstall, Jeremy: *The Fisherman: The Sociology of an Extreme Occupation*, McGibbon and Kee, London, 1969.

Untergasser, D.: *Handbook of Fish Diseases*, T.F.H. Publications, Neptune, NJ, 1989.

VanStone, J. W.: *Athapaskan Adaptations: Hunters and Fishermen of the Subarctic Forests*, Aldine, Chicago, 1974.

Villars, T.: *The Chemical Senses of Fishes*, Academic Press, London, 1974.

Ward, Barbara E.: *Varieties of the Conscious Model: Fishermen of South China*, Tavistock, London, 1965.

Warner, William W.: *Beautiful Swimmers: Watermen, Crabs and the Chesapeake Bay*, Little, Brown, Boston, 1976.

Index

E

F

G

❑❑❑